Evolution and a Creator?

To Elaine

Evolution and a Creator?

Graham Beale

© Graham Beale, 1991

All rights reserved. No part of this publication may be reproduced, stored in a retrieval system or transmitted in any form or by any means, electronic, mechanical, photocopying, recording or otherwise, without the prior permission of the Copyright holder.

Published by:

Clinical Press Limited,
Registered Office: Redland Green Farm,
Redland Green, Redland, Bristol, UK, BS6 7HF

British Library Cataloguing in Publication Data

Beale, Graham
    Evolution and a creator?
    1. Christian doctrine. Creation related to theories of evolution of organisms
    I. Title
    231.765

ISBN 1-85457-023-4

Typesetting from customer supplied disk by Santype International Limited, Salisbury, Wiltshire.

Printed by Albany House Ltd., Coleshill, Warwickshire.

Contents

Introduction

Acknowledgements

Chapter 1   The Biblical "Creation"

Chapter 2   Darwin's theory of evolution

Chapter 3   neoDarwinian evolution of life on earth

Chapter 4   Bones

Chapter 5   Bone

Chapter 6   Cells

Chapter 7   Life

Chapter 8   The evolution which has actually occurred

Chapter 9   Theories about the origin and the evolution of life on earth

Bibliography

Glossary

Introduction

This study started by chance with an injured kiwi. I was asked to assist with its assessment and care by Xraying it. This aroused my curiosity in kiwis and I began to study them with the expectation that I would find many features which I would not understand. Surprisingly, however, a number of the birds' features seemed to be similar to those which were present in humans, and many of the things which should have been quite strange to me seemed to be subtly understandable. As the study proceeded and the incomprehensible was being actively sought, I opened a kiwi's beak to examine its throat. Astonishingly, it looked remarkably like a human throat and it suddenly seemed that if I stopped seeking the inexplicable and studied only the understandable, it was possible to examine the various biological systems of the kiwi's body without being too confused. I could apply my medical knowledge to them and obtain a great amount of insight into their forms and functions. The same technique was then used to study the biological systems in other birds, mammals, reptiles and amphibians with the same result.

It then became evident that the understandable in one species was understandable in another species because they were similar to each other, and that within the similarities two groups of components were always present. There were those which were the same in each similar feature and those which were different.

The studying of the understandable was then widened to include all the kingdoms of life and any differences which were found in similar features were set aside for later analysis. They were being contemplated in a desultory way one day when it suddenly became apparent that the differences which were present in bones always occurred in the same limited number of ways. They did not occur illogically as they should have had they evolved by chance and subsequent natural selection. They occurred predictably. Shortly afterwards it also became clear that if the secret of life itself was to be found anywhere, as it is a living process, it must be possible to find it in the life which was around me. If it could not be found there, it certainly could not found elsewhere on earth nor could it be found in the past as it was no longer there.

At that stage it had become evident that if the Darwinian theory of evolution was as faulty as it seemed to be, an agency other than chance, or perhaps additional to it, must have been involved in evolution and perhaps in the origin of life itself. If that was to be proved, it could only be proved by studying the first spark life and establishing its origin. If it could be shown that some of the genes of that cell were permanent, chance would not then necessarily be an acceptable explanation for their appearance nor for the appearance of life itself. The study was then directed towards this end.

The first part of this book has been written to establish amongst other things, that at the time of the publication of Charles Darwin's book "The Origin of Species" in 1859, most scientists and all theologians believed implicitly in the Gospel Truth of the Bible and its account of the Creation. Only a few scientists, mainly geologists, questioned this and in the eyes of the believers, in doing so they committed blasphemy. An example of how unquestioning the view of the believers was, is encapsulated in William Paley's book "A View of the Evidences of Christianity". It was a standard, well-respected religious text of the first half of the 19th century with its 23rd and last edition being published in 1859. Darwin certainly read one of the earlier editions when he was at Cambridge and undoubtedly understood its two essential injunctions:

1 Study the evidences for and against Christianity with an open receptive mind and try and avoid prejudices and prejudgements

2 Because the Bible is completely true, it can be studied as such

Paley's display of logical reasoning based on these injunctions was excellent and explains why his book was so successful for so long, but his basic problem was that he disobeyed his own injunctions and made an erroneous prejudgement about the truth of the Bible. It took Darwin to prove that he had made this mistake and when it was unmasked, Paley's excellent logic collapsed into nothing. Darwin's unmasking of this fallacy about the Bible also sent its theory of the Creation crashing into oblivion.

The modern update of the Darwinian theory of evolution, the neoDarwinian theory, has now reached the same position of reverence in Science that the theory of Creation had in 1859. Almost all biologists accept that the neoDarwinian theory need no longer be regarded as a theory. Instead they accept that it is a biological Law which is the complete and only truth about the origin and the evolution of life on earth. Few question this and those who do so tend to be regarded by some of their contemporaries as being scientific heretics. Examples of the belief in the strength of the theory can be found in many books, and the following examples have been drawn from three quite different sources.

The first is by scientist who asked a computer to explain how randomly-occurring changes to genes had resulted in evolution. Richard Dawkins in his book 'The Blind Watchmaker" begins its preface by saying "This book is written in the conviction that our own existence once presented the greatest of all mysteries but that is a mystery no longer because it is solved. Darwin and Wallace solved it, though we shall continue to add footnotes to their solution for a while yet".

The second is by a biologist who has spent a considerable amount of his time searching the earth for fossils. Richard Leakey in the introduction to his book "The Illustrated 'Origin of Species' by Charles Darwin" says that "all aspects of modern evolutionary biology can be seen as part of a research programme inaugurated by 'The Origin of Species'. It is without doubt the most important biological work ever written".

The third is by a naturalist who has travelled the world studying the wonders of nature. Sir David Attenborough, in the first chapter of his book "Life on Earth" says "Since that time (the publication of 'The Origin of Species'), the theory of natural selection has been debated and tested, refined, qualfied and elaborated. Later discoveries about genetics, molecular biology, population dynamics and behaviour have given it new dimensions. It remains the key to our understanding of the natural world and it enables us to recognise that life has a long and continuous history during which organisms, both plant and animal, have changed generation by generation, as they have colonised all parts of the world".

These statements typify the prevailing view of Science about the neoDarwinian theory, but they are not held by all scientists. For example, in the introduction to his book "The Great Evolutionary Mystery", Gordon Taylor says that "Darwin's theory of evolution by natural selection, which has stood as the one great biological law comparable with the laws of physics for more than a century, is crumbling under attack. Biologists are discovering more and more features which it does not seem to be able to explain—-". He then goes on to examine the evidences for and against some of those features.

Stephen Gould in his books on natural history also believes that, even though he accepts the neoDarwinian theory, the fossil records do not show the gradual change which Darwin hoped would be found and which neoDarwinism needs to have found for it to prove that the theory is true. There is no evidence that the gradualism, an intrinsic component of neoDarwinism, has occurred. Instead, there is clear fossil evidence that species in the fossil records remain unchanged for millions of years then suddenly disappear and are replaced by new species which are substantially different to those which have vanished but which are clearly related to them.

The negative views which have been expressed about the neoDarwinian theory have not usually been made in conjunction with the advancing of replacement theories for it. They have simply been criticisms of the existing theory. Some however have taken a more positive stance and without necessarily wanting to abandon all aspects of the neoDarwinian theory, they do not accept that it is the only Truth about evolution. They believe that influences from space have also affected at least some of the evolution of life. For example, recently the astronomers Professor Chandra Wickramasinghe and Sir Fred Hoyle claimed that viruses come to us from time to time from outer space and if they don't influence our processes of evolution, they do at least cause some of our periodic epidemics. Although their belief ignores the essential nature of viruses and their relationships to cells, it does reflect the scepticism which some thoughtful scientists have about the neo-Darwinian theory of evolution.

This book carries that scepticism much further and does so on the basis of discoveries which have been gleaned from a variety of fields. If the facts which I have presented are wrong this can be proven and so-doing will cast into question the deductions and suggestions I make based upon those facts. However, if my facts are right then my opinions require serious consideration.

Graham Beale

Acknowledgements

It is with pleasure that I acknowledge the following photographers and organisations whose photographs have been included in this book:

Ross Clayton, University of Waikato, Hamilton, New Zealand, page 3
Aidan Dowle, page 2, 24, 32A, 66B
Forest Research Institute, Rotorua, New Zealand, page 66A
Lloyd Homer, D.S.I.R., Lower Hutt, New Zealand, page 18
National Library, Wellington, New Zealand, page 67
Adya Singh, page 22, 23, 52, 60
David Wilde, Meat Industry Research of New Zealand, Hamilton, New Zealand, page 68

I am particularly grateful to Rick Buchanan for photographs on pages 2, 32B, 33A&B and for his reproductions of the radiographs and the charts.

Chapter 1

The Biblical "Creation"

Before Charles Darwin published his book "The Origin of Species" in 1859, it was widely believed that the Bible was the true record of the Works of God and included in this belief was the acceptance that its account of the Creation of the heaven and earth and all its creatures was completely accurate. Understanding this, in the 17th century, an Irish Archbishop, James Ussher, established exactly when the Creation happened. By analysing the Bible's accounts of events such as the reigns of Kings and the date of the Exodus, he calculated that the Creation had occurred in 4004 B.C., and by refining his methods, other scholars determined that as part of the Creation, Adam was created on Saturday, October 3rd of that year at exactly 8pm. It seems silly now to think that people could accept that these statements were absolutely true, but they did so, and they did so because these were the pronouncements of holy men, and when holy men studied the Holy Scripture, how could they be wrong?

    The Bible also contained information about God's thinking and the way He wanted His world to develop. As explorations of the earth were continually demonstating, during the Creation, God had not created a fully-populated world. Instead, He had created examples of His creatures each in their own unique locations and had then bidden them to go forth, multiply and populate the earth. They were to continue this task until the earth was full of His creatures all living together in harmony, but by the middle of the 19th century, it was obvious that few species had so far achieved any great success in this direction. In fact, many had not yet even reached every corner of the earth, and there were good reasons for this. After the Creation, as each plant or animal had populated its own area and had spread into new ones, some had done so more effectively than others. Rapid breeders had populated new areas more rapidly than slower breeders, and the fast and the flighted had spread faster and further than the slow and the wingless. Most seemed to be continually trying to obey God's commandments in their own particular way despite the many obstacles posed by the mountains and the seas, but in some places some strange anomalies were found. In some instances, the slow had actually spread more effectively than the quick, and

2 EVOLUTION AND A CREATOR?

The Holy Bible
This particular Bible was printed in 1796. Using the religious knowledge of the time the printer annotated the margin of the Book of Genesis with the date of the Creation (4004 B.C.) and the dates of subsequent important Biblical events.

this implied that some creatures were not following God's wishes as assiduously as others. Despite this and the disappointment which this must have brought to God, there was no Biblical record of Him having been so concerned that He had felt it necessary to change any creatures after the Creation. Despite the foibles which some had manifested, when He created them He created them perfect and that was the way they would stay. They were the same now as they had been in the Beginning and they were the same now as they would always be. They were all permanent, unchanging and immutable.

It was accepted however that within the boundaries of this immutability, some minor variations could occur. For example, the colourings or the markings of individuals could vary slightly from individual to individual within the same species, without it causing concern, but if more significant differences were present, they were not present because the individuals were variations within a single species, they were present because the individuals belonged to different species. The many breeds of cattle illustrated this point. Although the blotches of colour in their hides could vary in size and location in different animals within the same species, where more significant differences were present such as differences in the forms of their horns or differences in the forms of their bodies, they were obviously different species which had been separately created.

The presence of the similarities which were evident between species were also explained. During the Creation, as God was creating new species, He used a number of basic patterns, changing them only slightly as He created the next new species. The Swedish botanist born Carl von Linne but later latinised to Carolus Linnaeus, recognised this and in 1737 published his study of some of the patterns (though he did not call them that) which he found were common to species and the differences which he also found. He used them to formulate his method of classifying species. He gave each species two names – one which highlighted one of its unique features, and the other which highlighted a pattern or a cluster of patterns which were common to it and other similar species. These similar species were then aggregated into groups which he called genera, and using further common patterns, he grouped similar genera into orders and similar orders into classes. His method of classification were so simple, so obvious and so successful, that it became adopted as the favoured methods of classifying species and as further creatures were discovered, the presence in them of patterns which were already recognised in existing genera, orders and classes, permitted them to be readily and correctly classified. Among the new species which needed to be classified were fossils.

Fossils are many things. Some are impressions of ancient actions of creatures such as footprints which persisted in mud until the mud set and

Fossil.
The shells of some animates (*Monitis richmondiana*) of 195Myrs ago which have been preserved in stone. It is little wonder that before the true nature of fossils was recognised, some thought that they were examples of God's artistic activities.

turned to stone thereby preserving them in perpetuity. Some are impressions of plants or animals which have been preserved in a similar way and although the original plants or animals have long since decayed and vanished, their impressions have remained faithfully preserved in stone. Some are parts of plants or animals which have been preserved in drops of tree gum which have then been preserved in stone or coal. Some are parts of plants or animals which have been preserved in ice and some are the shells or the bones of ancient animals. All are fossils but their true nature was not recognised until the late 18th century. Until then they were thought to be examples of God's artistry in the rocks or were the deceitful acts of the Devil. However in 1770, a Swiss naturalist, Charles Bonnet, suggested that bony fossils were actually the remains of dead creatures and as none of the fossil species were still alive on earth, they must be the remains of species which had become extinct.

Absolutely unthinkable cried the Theologians! Species could not be extinct! God had created creatures because that is what He knew He wanted in the world. No more and no less! Creatures could not become extinct because there was no way that that could happen and even if there was, there was no way that God could let it happen! What an outrageous thought! But – perhaps Bonnet was right. If fossils were actually the remains of creatures, because there were none of them alive on earth, they must be the remains of extinct species. How could this have happened? God could not have done it. Definitely not! Had He done so, it would be recorded in the Bible and there was nothing recorded there. It must have been the result of a natural disaster of such immense proportions that even God himself had not been able to control its ultimate fury and the Bible recorded just such an event!

It was the Great Flood. When it was coming, God knew that it was going to be so huge that it may endanger His Work and that some of His creations may not survive it. Therefore He instructed Noah to build an Ark, but despite Noah's best efforts, a major ecological disaster happened. When the Flood finally came and Noah called upon the animals to embark in his Ark, creation by creation, breeding pair by breeding pair, he had not bargained on the frivolous nature of some of them. Despite his impassioned pleas, when the Ark finally floated free on the rising flood waters, some were not on board, and when the flood waters finally engulfed the whole of the earth, they were drowned and were buried by the mud of the Flood. Whole species became extinct. They were not made extinct by the actions of God, indeed, God had tried to save them. They were made extinct by the combination of the Great Flood, Noah's error of judgement and their own foolishness. Gradually, as the great masses of flood mud turned to stone, within that stone, the engulfed species also turned to stone and thereby became fossils.

With the acceptance that fossils were the remains of extinct species, because they had all been God's creations, naturally their bodies contained many of the patterns of features which were present in living species. Naturally also, the Linnaean method of classification of species was applicable to them and by the beginning of the 19th century, Georges Cuvier was identifying even more widely distributed patterns and was grouping similar classes into phyla. His successor, Richard Owen, the last of the great Creationist anatomists, was still expanding the Linnaean method of classification and was grouping like phyla into kingdoms of life when Darwin's thesis was published.

Initial doubts about the Creation

With the publication of "The Origin of Species" biological thinking took a sudden step in a different direction and the biblical principles which underpinned the existing scientific thinking were simply abandoned. However, this dramatic shift was not entirely the result of Darwin's publication. Some scientists had already begun to worry about the validity of the Biblical account of the Creation and their studies had prepared the way for Darwin's revelations

There are two parts to the Bible's account of the Creation – the Creation of heaven and earth, its waters and the dry land, and the Creation of the creatures themselves. As the accuracy of the Biblical account of the Creation gradually came under suspicion, it was the account of the Creation of the earth which came to be questioned, not the account of the Creation of heaven or the account of the Creation of the creatures which inhabited the earth.

A geologist found that his observations about the earth could not be reconciled with Ussher's estimate of the date of the Creation. In 1785, James Hutton a Scottish naturalist, examined Hadrian's wall and compared the way that nature had changed the rocks in the wall with the way that it had changed the same types of rocks in the adjacent mountains. In 1785, the wall, known to have been built between 122 A.D. and 127 A.D., was over 1600 years old and Hutton reasoned that if nature had taken 1600 years to cause the trivial changes which he found in the rocks of the wall, it must have taken much much longer than 6000 years since Ussher's date for the Creation to have produced the changes which he observed in the mountains. There were so few changes in the rocks of Hadrian's wall and so many in the rocks of the mountains that it must have taken millions of years for them to have happened and if that were so, the earth must have been created much earlier than 4004 B.C.

Hutton's fortitude in publishing his thoughts has to be admired. By challenging Ussher and the accuracy of his calculations, Hutton was challenging the Bible and he could be right only if the Bible were wrong or Ussher's interpretations of its contents were wrong. But the holy man Ussher had produced his findings by examining the Holy Bible. Ussher must therefore be right and Hutton must therefore be wrong. *Quad erat demonstrandum.*

6 EVOLUTION AND A CREATOR?

Hadrian's Wall.

Yet by any standards, Hutton's scientific investigative techniques were excellent. He made simple field observations; drew some conclusions from them and attempted to explain them within the accepted scientific beliefs of the day – the biblical account of the Creation. When he was unable to do so, he then challenged the accepted beliefs about the Bible, but instead of his contemporaries reconfirming his observations and dispassionately examining his conclusions, because it could not even be remotely considered that the Bible was wrong, Hutton had to be wrong.

However, other evidence was found which questioned the accuracy of the Bible. In 1791, the English surveyor, William Smith, was digging a canal through a rocky area near the City of Bath and observed that the rock was layered. According to the geological views of the day, the layers of rock had been formed by the Great Flood mud but Smith reasoned that several layers of rock could not have all been laid down in the same Flood. Several layers meant several Floods, yet the Great Flood was the only Flood recorded in the Bible. Perhaps the Bible was not the complete record of all of the great natural disasters of the past that it was believed to be.

What a stupid thing to say, cried the critics! The Bible was the Gospel Truth and the record of all the major natural disasters of the past. If it was not in the Bible, it had not happened and Smith's thoughts crashed into the same wall of conservative thinking that Hutton's had hit. The Bible was right. Therefore Smith was wrong.

Increasing doubts about the Creation

Smith also wondered about the fossils which he saw in the rocks. In the deepest layers, the layers which he presumed were the oldest, he found simple fossils, and as he progressed upwards through the layers from the oldest to the youngest, he found that the fossils became progressively more complex. Were the fossils actually recording an ascent of life?

That was theologically totally unthinkable! All God's creations had been created during the Creation and only during the Creation. They were all unchangeable because they were all perfect and those which had become extinct, had all been made extinct by the one Great Flood and it alone! THE BIBLE WAS RIGHT!!! How could there have been an ascent of life? Stupid fellow!

But others made similar observations and were also not so sure about the Bible. In 1830, Charles Lyell published the first volume of his book "Principles of Geology" in which he detailed more evidence about the layering of rocks, reaffirmed Hutton's thinking on the age of the earth and produced further evidence of the presence of different fossil forms in different rock strata. In 1844, Robert Chambers asserted that the most simple fossils were regularly being found in the deepest and oldest layers of rocks and that fossils became increasingly more complex as one ascended through the layers. He believed that there had been a gradual ascent in the complexity of life with the passage of time.

The then Professor of Geology at Oxford voiced the anger which many felt about these kinds of radical thoughts. If Chambers was right "the labours of sober induction are in vain; religion is a lie; human law is a mass of folly and a base injustice; morality is moonshine; our labours for the black people of Africa were works of madness; and man and women are only better beasts" but his rantings were of no avail.

The Creation is swept aside

Despite the prevailing hostility to them, the thoughts of observant people were read and thought about by other thoughtful people and gradually their belief in the Bible's absolute infallibility was eroded. It was the observations and beliefs of Lyell particularly which helped Darwin as he tried to understand the things he saw, and the thoughts he thought during and after his five years on the H.M.S. *Beagle*.

Charles Robert Darwin was born in Shrewsbury, England on February 12th, 1809, and was educated at the Shrewsbury Grammar School. His first University was Edinburgh where the sights and sounds of early nineteenth-century medicine rapidly convinced him that he lacked the necessary fortitude for its study and he moved to Cambridge to study the classics and mathematics prior to entering the Church. Apart from his scholastic endeavours and his active pursuit of pleasure, Darwin also maintained his childhood interest in botany and in this respect he was lucky. He became friendly with the Reverend Professor Henslow, the Cambridge Professor of Botany who in 1831, after Darwin had graduated, encouraged him to apply for the post of naturalist on the H.M.S. *Beagle*. His luck also stayed with him when he was gathering together the small library of books which he took with him on the voyage. Included among them was the first volume of Charles Lyell's recently published "Principles of Geology" and although the book had been branded as rubbish, Darwin was able to read it, and the second volume when it appeared, away from the clamour of such prejudices. Although Darwin is known to have started the voyage contemptuous of the value of geology and, probably, a complete unbeliever in Lyell's work, by the end of the voyage, he had been able to assess for himself the validity of Lyell's concepts and had become a first-class geologist himself. By the end of the voyage, he was already moving towards deciding for himself that the Bible was not the geological truth about the origin of the earth, and he was also starting to reassess the observations he had made about God's creatures. However he still had some way to go before he could believe that he had actually stumbled on the real truth about the origin of life on earth and its subsequent evolution.

Naturally, they were very difficult conclusions to come to. He started the voyage believing that all God's creatures had been created perfect and he was continuing to study them within this context when the *Beagle* visited the Galapagos Islands. There he received a shock. He wrote:

> "I never dreamed that islands about 50 to 60 miles apart, most of them within sight of each other, formed of precisely the same rock placed under quite similar climate, (and) rising to nearly equal heights could be so differently populated (by different species)".

He was struck by the variations in finches on the different islands. Biblically, as all the finches on the various islands were descendents of the same created pair of finches, they should all have been the same, but they were not. They were obviously different species because they each had unique features, and they could only have them if they had descended from different creations. If the finches had responded to God's direction to go forth and populate the earth, as they could fly, the descendents of each of the different creations should have populated all the islands and all the islands should therefore have all the different species. But this was not so. Each island had its own unique species. Therefore, either each species had been created on each island and had not responded to God's directive or they were all descendents of one created pair and somehow they had changed since the Creation. Things on those islands were obviously not what they should have been but instead of closing his mind and saying that it was just another example of God acting in His mysterious ways, Darwin asked himself the then theologically impossible question – if they had changed, how had it happened?

He could not begin to answer that question immediately. In fact, it was a long time after the voyage had ended in 1836 before he knew enough to do so and he did so after he had read the writings of an Anglican clergyman named Thomas Robert Malthus. Between 1798 and 1830, Malthus published essays on the possible problems that human populations would encounter when God's wishes had been fulfilled and the earth was fully populated with all of His people. From his investigations of the population increase in many countries including North America, he found that the human population had doubled every 25 years, but food resources had not kept pace with this increase. In some instances, the food resources had taken 50 years to double, and if this trend continued, the earth would ultimately be unable to supply the people's needs. Although Malthus's writings dealt only with the effects of human overpopulation and the problems of limited food supplies, when Darwin read this he perceived that there was an idea which was applicable to all forms of life. Darwin already knew that the number of offspring which parents produced always exceeded the number of parents but despite this, species tended to maintain stable populations generation after generation. This meant that many offspring died before they could breed and as all offspring were always competing for the same food supply, those which obtained more than their fair share were ultimately the ones which survived long enough to propagate the next generation of the species. While some were surviving, thriving and propagating, most were dying, and it was the subtle differences which were present in some individuals which gave them the advantage they needed to survive, thrive and breed. If these advantages were inherited by their offspring and they accumulated in descendent lineages of offspring, gradually the original species would change and in time a new and better species would appear. He had cracked the secret of evolution! He wrote:

> "It at once struck me that under these circumstances, favourable variations would be tend to be preserved, and unfavourable ones be destroyed. The result of this would be the formation of new species. Here then I had at last got a theory by which to work".

It was several years before Darwin could finally look beyond these completely sacriligious views and write:

> "I am almost convinced (quite contrary to the opinion that I started with) that species are not (it is like confessing to murder) immutable. Heaven forsend me, from Lamarck's nonsense of 'a tendency to progression', 'adaptation from slow willing animals', etc! but the conclusions I am led to are not widely different from his, though the means of change are wholly so".

But Darwin had still to come to terms with his conscience. He knew and understood what had happened to life on earth and he knew and understood that it was at complete variance with what was written in the Bible. He also observed that some and perhaps all species could change and were not as perfect and unchanging as they were believed to be, and he understood that fossils recorded that there had been an ascent of life. He had reasoned that by reversing the ascent and looking backwards in time, he could define when life began. He realised that the further one went back in time, the more simple the fossils became and if one went back sufficiently far, one must ultimately come to the most simple of all life forms, the prototype life form. Life must have started in that prototype and, regardless of how it or life appeared, all life must have evolved from it and have done so without God's assistance. None had been created.

He recognised all this but either would not or could not write it down in book form. Perhaps he was too uncomfortable with the magnitude of it all. Instead, he spent eight years investigating and writing about barnacles and other things, and it was not until another scientist came to the same conclusions and was preparing to publish them, that Darwin was forced to act. To establish scientifically that his views were his own and had not been pirated, he prepared a short paper for presentation to the Linnean Society meeting of July 1st, 1858, but could not present it himself. A family illness and/or his mental ferment about his ideas, prevented him from doing so and a friend read his paper for him. Alfred Russel Wallace also had his very similar paper read to the meeting and the resulting impact of their papers was – nil. Despite the nature of the society and its receptiveness to new ideas, its members were also christians and as such they believed in the Gospel Truth of the Bible. Therefore with christian charity, they could tolerantly listen to and politely ignore revolutionary thinkers who tried to overturn that belief, and this they did. The two papers were then published in the *Journal of the Linnean Society* on 20th August, 1858, and again they had so little impact that in his 1858 annual address, the President of the society observed that the year had brought forth no papers of any distinction. The passing of time has revealed how delightfully he understated the situation.

With the presentation of his paper, Darwin's activities took on an air of urgency and he immersed himself in the preparation of his book. Fortunately, in his publishing company, he found an organisation with the marketing skills which were necessary to ensure that the book was success. By careful advanced publicity, the public's appetite for the forthcoming publication became so whetted that on the day of its release on November 22nd 1859, the 1192 copies of the book which were available for sale, sold out immediately and the "The Origin of Species" was an instant sensation. With the few strokes of his pen, Darwin completely undermined people's deep and inate belief in the absolute truth of the Bible. He proved that life on earth had evolved and had not been created and he also announced that except perhaps for some involvement in the appearance of the very first life form, God had played no part in the appearance of Man or in the appearance of any other creature which had ever lived on earth.

The theological world was absolutely stunned by these anti-religious messages but Darwin's many observations were so patently true and so clearly explained by his theory of evolution, that rapid massive scientific endorsement of his theory was inevitable. Public endorsement however took a little longer. Darwin's claims were so unusual and provocative that the public could not cope with them in their entirety and found it easiest to grapple with the one aspect of his theory which affected them most. Where had they themselves come from? Had they descended from the apes? As the theory said that more complex species had descended from more simple ones, as apes were our nearest relatives and were more simple than us, we must have arisen from them. That was the natural conclusion of Darwin's theory and around this point, some very virulent arguments raged. In this respect, Darwin was fortunate that Thomas Huxley, an influential scientist, became his very articulate champion allowing him to stay out of the main stream of the debates and abuses. Huxley had very rapidly grasped the concept of Darwin's theory, so quickly in fact, that when he first read the book, he said "How stupid not to have thought of that", and on 30th June, 1860, seven months after the publication of the theory, Huxley took part in the most important of the public debates on the controversy. The debate was between Science in the form of Thomas Huxley, and the Church in the form of Bishop Samuel Wilberforce, the Bishop of Oxford. The problems of Man and his descent from the apes formed its central theme and as the rhetoric volleyed and thundered, the more eloquent tongue of Huxley was victorious; Science won and the Church's opposition was completely blunted. From then on, more and more scientists and churchmen came to accept the strength of Darwin's theory until finally, by the time he died in 1889, the opposition to him had become so muted, that he was buried as a national hero in Westminister Abbey, in a bastion of the Church his theory had so savagely mauled.

Chapter 2

Darwin's theory of evolution

The theory which Darwin developed to explain the origin and evolution of species was simple and elegant. Life had not been created. It had evolved and had done so in a very simple way. There were always minor differences present between individuals of the same species. Some of these improved the chances of some of the individuals with them surviving and propagating their species in preference to their contemporaries, and if that happened, nature naturally selected them to do so. As these advantageous differences in characteristics progressively accumulated in lineages of descendants, new and more complex species gradually appeared. This progressive appearance of new and more complex species was evolution. As part of the process, innumerable intermediate life forms also appeared but as evolution continued to progress, the developing new species continually disadvantaged both the intermediate life forms and the already existing older more simple species and as a result, they died out and became extinct. Extinction of species was therefore a natural consequence of the evolution of species.

Fossil finds confirmed that species had become extinct and they also explained how life had actually started on earth. As one retreated back in time the fossils became progressively more and more simple and it was logical that if one went back sufficiently far, one would ultimately reach the most simple of all life forms, the prototype life form. Darwin was unable to determine how it had actually come into being and believed that God breathed life into it, but once it had come into being, all subsequent life evolved from it by the processes of natural selection.

The adoption of his theory naturally caused earlier observations and explanations about life to be re-evaluated. Suddenly, the shifting locations of the patches of colour on cows' hides became totally uninteresting. They were obviously small variations which were always occurring and if at some time one particular form of patch improved the chances of that cow thriving and breeding in preference to other cows, it would naturally be selected to do. Arguments also ceased about things such as whether birds with slightly different feather colourings were separate creations or were varieties of the same creation. They were obviously not creations at all as nothing had been created. They were all variations of the same species and the slight colour differences in their feathers had the same significance in them as the variations in the patches of colour in their hides had to cows. The presence of extinct species in different rock strata became much more readily explained. They were not species which had been been destroyed by the Great Flood. They were species which had become extinct as new and more complex species had evolved and had disadvantaged them. Nor were the different rock strata layers of flood muds. Each stratum had formed as the result of the continually-changing face of the earth and its seas and the extinct species which were fossilised in each stratum were species which had been alive on earth at the time that that stratum had formed. As had been suggested by Smith and others, the increasing complexity of the fossils in the progressively younger and younger rock strata was a true record of the ascent of life on earth.

Naturally, the Linnean classification of species was also reviewed, but rather than becoming invalid with the change from Creationism to Darwinism, it flourished because it was based on verifiable field observations and not on theology. Only the explanation for the presence of common patterns in species and the differences between them changed. The differences between species were variations which had been naturally selected as the species continued to evolve and the common patterns which were present in similar species were present in them because they had inherited them from their common ancestors. Naturally, some evolutionary changes had occurred to the patterns with the passage of time, but their basic elements continued to persist because of their advantages to the various species. The differences were evidences of evolution at work and the similarities were family resemblances which stemmed from common ancestors. With this new explanation in place, the Linnean method of classification flourished. It is still the method of choice for classifying species today.

Despite the apparent simplicity of these and similar redefinitions, they were not completely understood by all. The conservatives who were very reluctant to abandon the Creation, attempted to savage some of them by asking awkward questions while those who genuinely wanted to understand the theory, asked equally awkward questions. For example, it was not understood how an organ such as the liver, could be present in so many different flighted, land-based and aquatic species. Undoubtedly the liver had unique features in some species, but essentially, it seemed to be the same organ wherever it was found. Was it actually the same organ and had it been inherited from a common ancestor? If so, how had the ancestor evolved the liver in the first place and why had the liver not changed as the ancestor's lineages of descendants evolved into the many different species which had livers? Intense intellectual activities surrounded the attempts which were made to answer these and similar questions, and as the main source of wisdom, Darwin attempted to explain them in five further editions of his book. Initially, he thought that although most life had evolved in the way he postulated, he also thought that some evolution may have happened in a different way, but as time passed, he dropped the other suppositions and came to believe that life had evolved in one only way – by the natural selection of advantageous variations and their gradual accumulation in descendent generations. When he was requested to provide appropiate proof for this belief, he could only answer that although enough fossils had been found to prove that life had evolved and prove that some life forms had become extinct, not enough had yet been found to prove that evolution had happened in the way he knew it had. Life was and always would be in a state of evolution and was presently affecting all species which were alive on earth, but it occurred too slowly to be perceived. It took time for new species to appear, but regardless of how much time each species may have needed in the past to evolve, there had always been adequate time for them to do so as the world was certainly over two hundred millions years old. The discovery of more of the many fossils which were in the undiscovered segment of the fossil records would ultimately prove his theory, especially if the discoveries included a few of the innumerable intermediate life forms which were in the fossil records. They were in the fossil records. That he knew for certain, and and when they were found they would validate his theory. In the meantime, people would have to trust him and accept that life on earth had evolved in the way he said it had.

Others decided that organs such as the liver were not actually the same at all. Instead of them having come from a single ancestor, they were lookalikes, and in many instances, very good lookalikes indeed. Some things had been such good evolutionary ideas that several different ancestoral species had followed similar evolutionary paths and evolved similar types of organs. In the end, they had done this so successfully that the results had converged to the extent that they had become exactly the same in unrelated species.

But questions continued to be asked and continued to remain unanswered, and the most important ones were – how did the small variations which were essential to the evolution of new species appear in individuals in the first place? How were they then inherited by their offspring? Darwin was still grappling with these problems when he died, but before that, the work of Gregor Johann Mendel had already started to provide some of the answers. In 1866, seven years after Darwin's book was published, Mendel published his work on 'particulate inheritance'. He found that small changes would appear in the flowers of plants he was breeding in his garden and these changes would then continue to appear in subsequent generations of descendants of the changed flowers and the seeds which they formed. He reasoned that this could only happen if some particulate elements in the plants had changed and these had then been inherited in their changed forms by direct descendents. Darwin knew of Mendel's work before he died, but neither of them could relate his work to that of the other. Indeed, it seemed for a time that their beliefs were in direct conflict and this lead to Darwin's theory going into a decline for some time. Decades later however, after further biological discoveries had been made, their two theories and some of the subsequent discoveries, were blended together into the new Darwinian theory of evolution.

In 1909, Mendel's "particulate elements" were renamed "genes" by Wilhelm Johannsen and in 1953, James Watson and Francis Crick analysed them. They found that they were protein molecules of desoxyribonucleic acid – DNA, and established that they contained the genetic codes of life. Later, some genes were artificially changed by genetic engineers and the changes which occurred to the characteristics controlled by them, subsequently appeared in changed forms in the descendants of the genetically-manipulated individuals. These experiments proved that all characteristics and features of life are controlled by genes and by them alone and that no changes can occur to any characteristics or features unless their controlling genes change.

The way that they have changed in nature in the past was established. They changed by chance as they were being inherited by offspring, and when changes occurred, as they were always doing, if any of them improved the chances of a particular individual propagating its species, that individual was naturally selected to do so and the changed genes then persisted in its subsequent descendent lineages.

The origin of life was established. Although Darwin had only been able to guess at the way that it had come into being, once it was discovered that life was a feature of cells and them alone, it became evident that Darwin's prototype life form must have been a living cell. But instead of God creating it, it appeared by chance.

The various discoveries and opinions are not now normally grouped together into a formal theory in the way that Darwin's original theory was. Instead, the new theory of evolution, the neoDarwinian theory has become a modern overview of Darwin's theory, restating it in the context of subsequent biological discoveries. It is now agreed that all evolution has occurred as genes have changed randomly and where the associated changes to the characteristics or features controlled by the changed genes have improved the chances of the offspring surviving and propagating the species in preference to its unchanged contemporaries, it has been naturally selected to do so. The gradual accumulation of these advantageous genes in descendent lineages, have resulted in the progressive appearance of new and more complex species and as they have appeared, because evolution has been a continous process, many intermediate species have also appeared. As new and more complex species have evolved, they have also disadvantaged both the associated intermediate life forms and existing more simple species many of which have become extinct. The actions of chance which have caused all life to evolve, also brought life into being. At some time in the past, chance aggregated the right combination of molecules and the combination formed the ancestral cell of all subsequent life.

The strength of the neoDarwinian view of evolution is now such that it is no longer regarded as just being a theory. It is accepted by many as being a proven biological Law which is capable of explaining all observations which can be made about past and present life. The only problem with this however is that so far, no fossil evidence has been found which actually confirms it. Fossils certainly confirm that life has evolved and certainly confirm that species have become extinct with the passage of time, but so far, none has yet been found which prove that life has evolved in the way which is implicit in the neoDarwinian theory. Only about 100,000 of the reputedly many millions of extinct species which are believed to have existed have been found and none of them have been any of the innumerable intermediate life forms which are also believed to be present in the fossil records. As a result, the lack of the right kind of fossil evidence is still being rationalised in a way which is very similar to the way that Darwin rationalised it. The theory is right and the fossils which will confirm this are still in the undiscovered segment of the fossil records. When they are discovered, as surely they will be, they will prove that the theory is true but in the meantime, the genetic engineers have provided enough evidence to prove that the neoDarwinian theory of evolution is the truth which it is believed to be.

Chapter 3

neoDarwinian evolution of life on earth

With an understanding of the neoDarwinian theory of evolution, it is now possible to discuss some of the details which are believed to have been associated with the appearance of life on earth and its subsequent evolution.

The appearance of the first living cell

At its inception about 4700Myrs ago, the earth is believed to have been a whirling, cooling mass of incandescent gases which gradually liquified, then solidified, and became the earth. By about 4100Myrs ago the first solid rocks had appeared and by about 3500Myrs ago at least part of the earth had cooled sufficiently for life to survive, for about that time, it is thought that life appeared on earth.

The environmental conditions which prevailed then are believed to have been distinctly different to those of today and studies of volcanic rocks which formed then have revealed the nature of the differences. As they were erupted in their molten form and came into contact with the "atmosphere", the chemistry of the molten rock changed and these changes record the nature of the atmosphere of the time. The rocks show that the young earth's atmosphere lacked oxygen and contained significant concentrations of other gases including carbon dioxide. Rain storms continually washed these gases down on to the surface of the planet where they mixed with molecules which were being leached out of the rocks, and gradually the waters of the earth became charged with complex mixtures of natural chemicals. Simultaneously, a much greater percentage of the sun's energy, particulary ultraviolet energy, reached the earth than reaches it now, as any ionosphere which was present then lacked ozone. The combination of intense ultraviolet radiation and numerous electrical storms, provided enough energy to cause numerous fusions to occur between the various elements which were in solution and in this way it is thought that simple amino acids formed which then fused with each other progressively forming more and more complex organic molecules. As they accumulated in what

18 EVOLUTION AND A CREATOR?

An erupting volcano.
As the erupted molten rock cools and solidifies, its chemistry is modified by the gases it is in contact with thereby recording their nature. Some of the gases will be gases which have just been erupted by the volcano but others will be the gases which are present in the surrounding atmosphere.

have been called "biological soups", some molecules began to appear which successfully initiated and completed their own reproduction. Some of these then successfully came to assert dominance over the activities of other molecular collections and in the process of doing so, became rudimentary genes. As they and their slave molecules continued to stay in contact, some of the slave molecules came to assume the specific responsibility for keeping the clusters of molecules together and formed rudimentary retaining walls.

So far, no molecular complexes were alive in the sense which we now understand life to be as, regardless of the interactivities of various molecules, no molecular complex had achieved the security which a complete outer casement could provide. At any moment, molecular collections could and presumably did break up and drift apart, but finally, in one drop of soup, the right combination of molecules appeared and a living cell formed. A cluster of molecules which was dominated and controlled by self-replicating molecules, became completely surrounded by an outer wall of molecules forming a complete cell which was suddenly alive. However, before it could live successfully and become our ancestral cell, it still had to perform another most important function. It had to initiate successfully and complete successfully its division into the next generation of living cells for had it not done so, life would not have continued on earth. The cell successfully did this thereby spawning the next generation of living cells. With this process completed, everlasting life was established on earth.

The particular molecular combination which formed that cell is not held to have been the only living combination of molecules which ever appeared on earth, nor is it held that it was the only original cell which ever lived successfully and formed a descendent generation of living cells. But it is held, that because all living cells today share the same family resemblences, only one living cell was the stem cell from which all subsequent life on earth as we know it has evolved. If any other stem cell did spontaneously appear and did engender descendants, no evidence of them has ever been discovered.

Subsequent evolution

With the successful appearance of the first cell, its immediate descendants became the first species of organisms and evolution began to proceed.

As all of today's living cells use energy in the form of adenosine triphosphate – ATP for their processes of living and produce it by breaking down sugars, it is thought that the first major evolutionary change which occurred in cells was the appearance of anaerobic glycolysis. Initially, early cells are believed to have obtained their energy needs more from the spontaneous breakdown of molecules in the biological soups than from ultraviolet rays, but as cell numbers increased and the concentration of unstable molecules in the soups began to fall, sugar molecules are presumed to have remained plentiful. Changes which facilitated the active breaking down of these sugars into ATP, occurred in the genes of some cells and gradually these and similar genes accumulated establishing the biological processes of gycolysis. The process must initially have been anaerobic as no oxygen was available either in the atmosphere or dissolved in the soups for the cells to use. Anaerobic glycolysis was, and still is, a rather inefficient reaction and when atmospheric oxygen began to become available and began to dissolve in the soups, an evolutionary modification occurred which resulted in the more efficient processes of aerobic glycolysis coming into being. But before this, another major evolutionary improvement in cell metabolism had to occur – the evolution of photosynthesis.

As supplies of every form of unstable molecules in the soups were gradually used up by the rising population of cells, chance genetic changes gave some of the cells the opportunity to use a new source of energy. They progressively switched from using the declining sugar resources of the soups to manufacturing their own sugar supplies. Fossil and geological records show when this change occurred. They show that by 2700Myrs ago, a species of single-celled organisms, the blue-green algae or cyanobacteria were present in the waters of the earth and were using water, carbon dioxide and sunlight to manufacture sugars. Had the production of sugars been the only result of their photosynthetic activities, the evolution of life on earth would have taken a rather different direction to the one which it did, but their photosyntheses did more than that. Not only did photosynthesis form sugars, it also formed OXYGEN. The cyanobacteria continually produced more oxygen than they needed for their own living purposes and the excess passed out of the cells and accumulated in the atmosphere. There, some changed of it into ozone which gradually accumulated in the upper atmosphere forming a protective barrier against ultraviolet rays. With this, the flow of ultraviolet energy to the biological soups began to diminish, the earth environment began to change, and gradually the conditions which had originally permitted life to appear on earth vanished for ever.

Subsequent life on earth

The Multicelled Species

As time passed, other changes occurred to single-celled organisms and after about 2500Myrs of life on earth, perhaps 50 to 60 different species had evolved. There were so few of them and the individual organism of each species were so small, that had some of their representatives gathered together to muse on how their descendants would evolve in the future and how many different species they would form, one small drop of water would have provided them with an oversized conference venue. Yet despite their microscopic size, because of the immense amount of time that they had been at work and the immense number of them which had been involved, they had irrevocably changed the environment of the earth. Some of them had also managed to change their internal and external forms. By about 1000Myrs ago (variously thought to be as recently as 700Myrs ago or as early as 1400Myrs ago), apart from anaerobic glycolysis being a facility which was present in all species, aerobic glycolysis was present in most of them and photosynthesis was present in some of them. Some had evolved into nucleated species with their genetic material aggregated into a nucleus and some of these had evolved into species which propagated sexually. Some had also achieved mobility by evolving a flagellum and some had even managed to achieve mixtures of several or all of these features. The changes which had ocurred since the appearance of the first living cell were so great, that even though perhaps only 50 to 60 different species had ever lived on earth, modern biologists believe that by 1000Myrs ago, they could no longer be classified as being in one kingdom of life. They formed two. One was the new kingdom of eukaryotes which possessed a nucleus, and the other was the older kingdom of the prokaryotes which lacked a nucleus. Both kingdoms are still well represented on earth today in the Monera and the Protoctista kingdoms.

Although it is not possible to determine what evolutionary steps occurred to form the species with these distinctive biological features, it is accepted that each was evolved by neoDarwinian processes, with advantageous randomly-appearing genetic changes accumulating in descendent lineages. However, in the species which achieved mixtures of some of the features, it has been suggested that some may have obtained them in a non-Darwinian way. Some organisms may have engulfed others, with engulfed individuals managing to continue to live inside their predators. As the predators propagated, so also did the engulfed individuals and in some instances, the genetic materials of the two organisms merged and the combined features of both organisms then formed a new species. But regardless of how they formed, by about 1000Myrs ago, the days of single-celled organisms being the only kingdoms of life on earth, were almost over, and the days of three new more complex kingdoms of species were beginning to dawn. The multi-celled ancestral species of the fungal, plant and animate kingdoms came to be formed as clusters of single-celled organisms bonded together into new composite organisms using the special features which each of the single-celled organisms brought to the partnerships.

The fungi came into being as some single-celled organisms merged their cell walls and formed a string of composite cells which lived and propagated as a single organism. The only bonding which existed between the participating cells was the union of their cell walls, and using this mechanism, fungi began their successful colonisation of the earth.

In other cell clusters, instead of the walls of individual cells merging to form composite cells, the cells remained as separate entities glued to their neighbours by special glues which were produced by some of the cells. The glues had two separate forms. One group were protein-based glues and the

22 EVOLUTION AND A CREATOR?

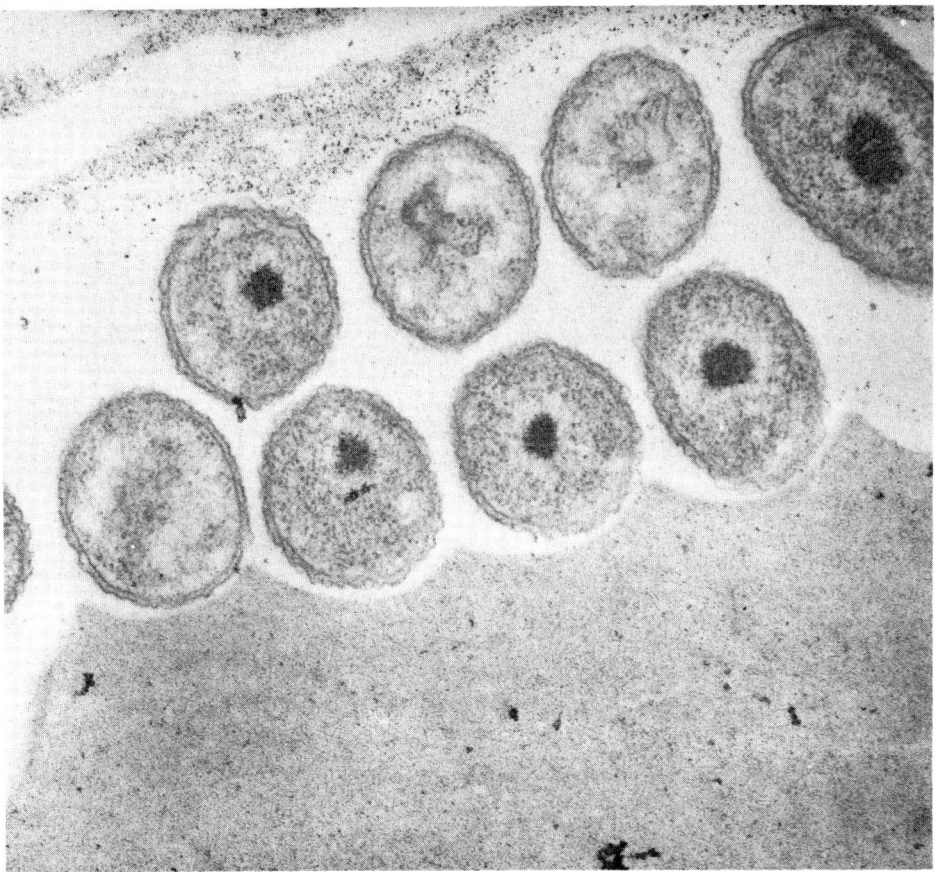

Prokaryotes of the Monera kingdom
Non-nucleated single-celled organisms. In this instance, they are sausage-shaped wood-eating bacteria which have been transected during preparation of wood for electron microscopic study. The crenated cell membranes form the complete outer boundary of each bacteria and their DNA is loosely concentrated in the centre of each organism. Their energy-producing and energy-using mechanisms are present in the surrounding cell fluid. The bacteria obtain their nutrition by passing wood-dissolving chemicals out through their cell walls. They then absorb the results. (magnification x 80,000)

organisms which they bonded together became the ancestral species of the animate kingdom. The other were carbohydrate-based glues and they bonded together the ancestral species of the plant kingdom. Both species included cells with most of the specialised facilities which had been evolved earlier – mobility, the aggregation of genetic material into nuclei and the ability to propagate sexually, but the power to photosynthesise remained selectively located within the ancestral species of the plant kingdom.

As the clusters of cells gave up their individual free-living life for a corporate life style, some cells evolved further functions which improved the living of the whole organism. In the animates, some cells formed collagen fibres – protein-based cellular excretions, which intermeshed within the cellular glue forming the structural support system of the animate species.

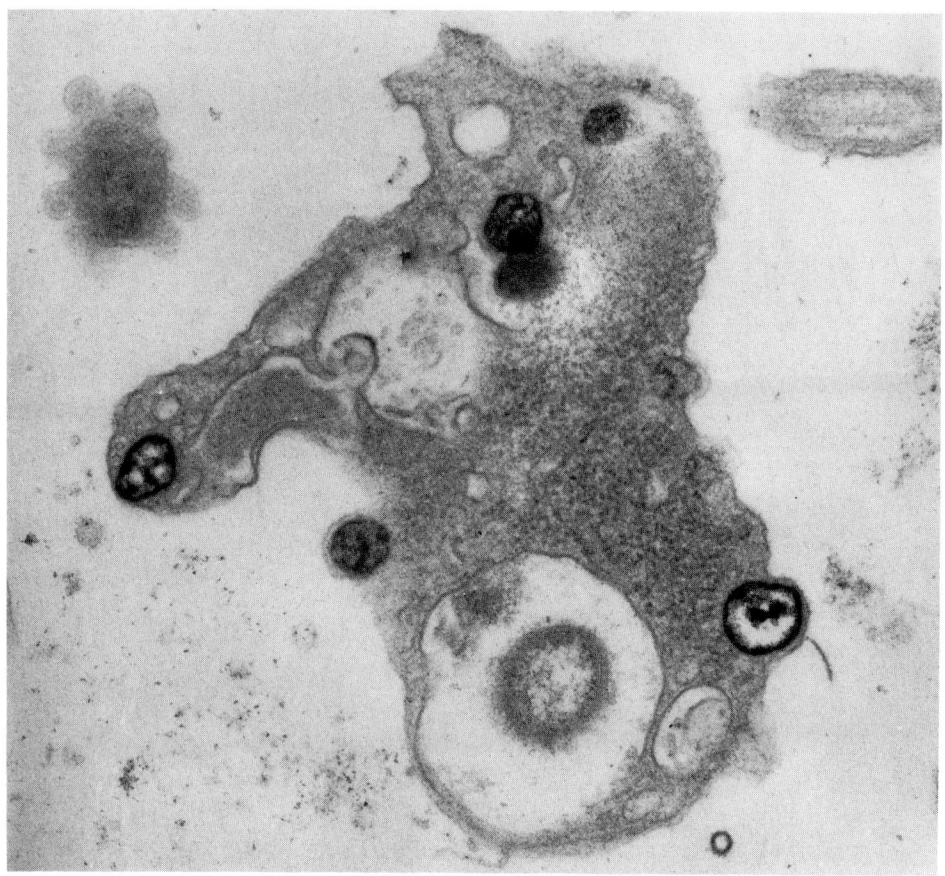

A eukaryote of the Protoctista kingdom.
An electron photomicrograph of a protozoon. Its form is similar to that of prokaryotes but it is a much more complex organism. Its DNA is concentrated in a properly constituted nucleus and its flagellum and the continual changing form of its cell wall give it its mobility. Apart from gaining some of its nutriment in the way that prokaryotes do, it is also carnivorous, and within its boundaries there are several prokaryote bacillae which have been engulfed and are being digested. (magnification x 16,000)

In plants, carbohydrate-based fibres performed a similar function. In each of the multicelled kingdoms, the fundamental differences which were present between the kingdoms were established at their outset. The fungi used cell membrane fusion to bind their cells together; the animates used protein-based glues and protein-based fibres to bind their cells together, and the plants used carbohydrate-based glues and carbohydrate-based fibres for the same purpose. The same still applies in the three kingdoms today. No living species have yet been found which uses mixtures of both types of cellular adhesives and intercellular fibres to bind their organisms together.

Although the ancestral fungal, plant and animate species may not have appeared contemporaneously, once each came into being, the evolution of their kingdoms proceeded rapidly, accelerated by their abilities to reproduce

Fungi.
These fungi are well organised multicelled organisms. Although some fungi have some kinds of fibres which help support their form and their weight, essentially, they are held together only by the bondings between their cell walls.

sexually. Sexual reproduction gave them significant advantages over non-sexually reproducing species. During the propagation of non-sexually propagated cells, a dividing cell could only form a limited number of descendants and in the process of doing so it disappeared into them. Its genetic and other cellular material was split between them and was then brought to its full measure by the new cells' own actions. No genetic material came into any of the descendent organisms from any outside source and the chances for genetic variations to occur were therefore limited. The rate of evolution in these species was slow. However, when propagation occurred sexually, parent organisms could prepare appropiate cells without themselves necessarily disappearing into the new cells. Consequently, the number of possible descendants which could be spawned rose significantly, and as sexually-prepared cells had to unite with other sexually-prepared cells to form new individuals, these unions meant that the genetic material in them was always a mixture of material from two different sources. Consequently, the opportunity for genetic changes to occur rose immensely and new species evolved rapidly.

As the rate of evolution accelerated in the multicelled kingdoms, the different methods of cellular bonding and the flexibility of the fibrous meshworks dictated the direction of their evolution. The absence of strengthening fibres in the fungi prevented them from evolving the same kinds of well-cordinated clusters of cells that the plants and animates evolved. In the animate kingdom, the pliable glues and the supple collagen fibres permitted the evolution of species with flexibility and mobility, and in the plant kingdom the relative inflexibility of their fibres and glues dictated their direction of evolution. Regardless of their ability to move or to stay fixed, both the animates and the plants progressively increased the numbers of cells in individuals and their success in this direction was considerable. New species appeared with more and more cells and a few species ultimately came to have individuals which were composed of tens of millions of individual cells all working together in harmony.

However, before such species could come into existence, both kingdoms had to evolve support systems which protected the cell masses from disintegrating under their own weight and in plants, some of the fibrous meshes stiffened and provided internal supports of the forms which are found in trees, while in the animates, a number of different calcium-based support systems appeared. Most of the animate systems were external systems such as the shells of shellfish, the armour of crabs and the formations of corals, but in some animates, the most important of all the support systems, the internal support system of bone, appeared. Regardless of the nature of the various structures that the systems formed and the kingdom of life which was involved, they were all the products of the activities of cells, rather than being cells themselves. This meant that although the cells which formed the structures decayed and vanished soon after death, some of the structures the cells had formed did not disappear quickly and persisted long enough for them to fossilise. Prior to the appearance of the support systems, the number of fossils which formed was extremely small because the cells which formed organisms decayed and disappeared too quickly for them to fossilise, but once the various support systems of the plants and animates appeared, a continuing sequence of fossils began to form and the fossil records as we understand them, started to come into being.

That was before 600Myrs ago and fossils of that time show that the life forms which lived then were simple plants and simple invertebrate animates. The fossil records then show that as time passed, more complex species progressively appeared and although this process appears to have occurred in each of the three multicelled kingdoms, it is best recorded and is best studied by examining the fossils which formed in the animate kingdom. Fossils show that by about 600 to 550Myrs ago, vertebrate fish had appeared with a bony skull, a bony axial vertebral column formed around a notochord and fins which were internally-supported by bones. During the next 100Myrs approximately, an increasing array of more complex bony fish came into being and amongst them was one which gave rise to a completely new species, the amphibians. Their body forms were different to those of fish and although they retained similar bony skulls and similar axial bony vertebral columns, instead of them having fins, they had paired forelimbs each with a five-fingered hand, and paired hind limbs each with a five-toed foot. With the increased mobility that this skeletal arrangement gave the amphibians, they began to colonise the land.

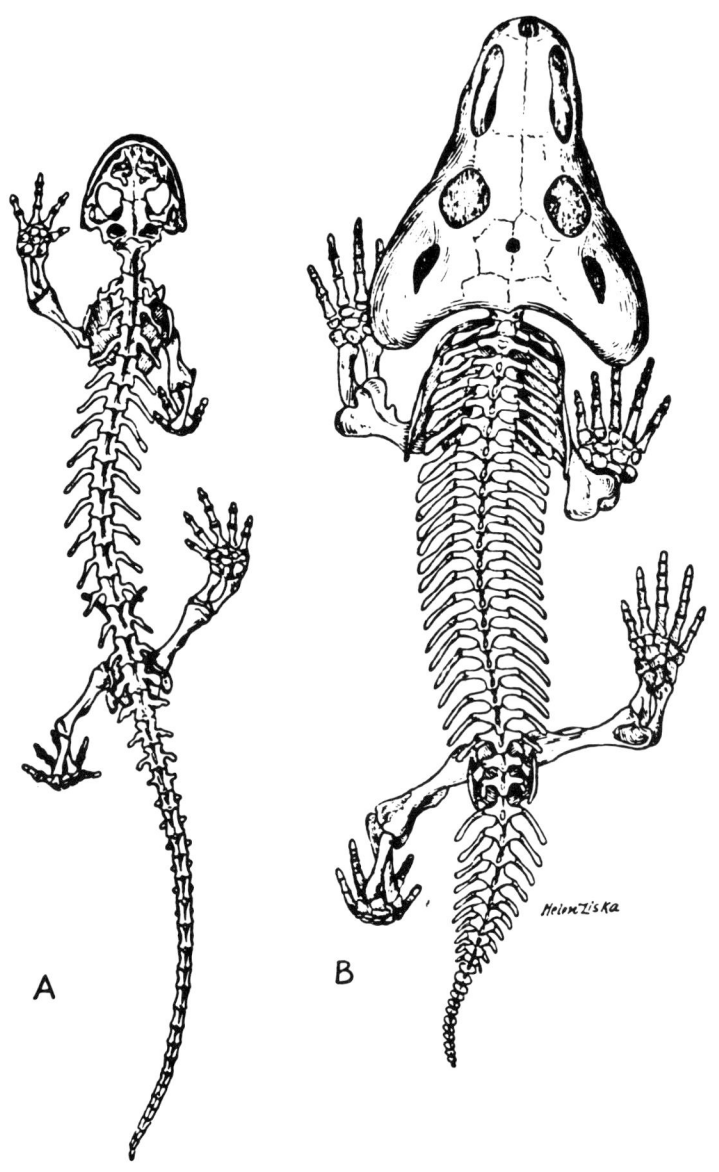

A modern pentadactyl tetrapodal vertebrate (PTV) *(left)* and a fossil amphibian (labyrinthodont) from the Permian era approximately 300Myrs ago.

What is not known is what relationship if any, existed between the evolution of the plants and the animates during the 50Myrs of the Devonian geological time of between 425Myrs and 375Myrs ago as huge evolutionary changes occurred simultaneously in both kingdoms. Plants from the sea began to colonise the earth and from them, trees appeared and came to reach their present canopy height. Insects came to populate the forests, amphibians came to populate the land and by the end of the Devonian, the first of the reptiles had evolved from some amphibians. Over the next nearly 200Myrs, reptiles rather than amphibians became and remained the dominant forms of animate life of the earth, but by about 180Myrs ago, they had begun to decline and the first birds and mammals had appeared. They then gradually became the dominant animate species of the land and the air while in the seas, the bony fish continued on their dominant way.

Many species of all of the five kingdoms of life are alive on earth today and their numbers give some idea of the success of the evolution of their ancestors during the last 1000Myrs. There are approximately 1.5 million different species alive on earth today and these include about 75,000 fungal and single-celled species; 325,000 plant species and 1.1 million animate species. The animates include 40,000 fish and tetrapodal vertebrate species, and the balance are invertebrates including insects. Many of the 1.5 million species have been identified thus making the estimate of their total reasonably accurate, but regardless of how many species there are alive on earth today, it is believed that it does not in any way to reflect the total number of species which have lived in the past. Estimates of that number are based on the neoDarwinian belief that creeping evolution has continually occurred and they range from as few as 50 million different species to as many as 4,000 million with perhaps 500 million being a reasonable compromise figure. These estimates do not include the innumerable intermediate life forms which are also believed to have appeared as each new species evolved. Though many of the soft-bodied life forms have never fossilised to the extent that they or their intermediate forms will ever be discovered, when those which have fossilised are summated with the fossils of their intermediate forms, there should be many many millions of different fossils still waiting in the undiscovered segment of the fossil records. Even if only a relatively minor percentage of them are ever found, some will have to be intermediate species, and should some of them be discovered, their discovery will provide the fossil proof which Darwin so badly wanted to prove his theory and which is still needed to validate the neoDarwinian theory. So far however, only about 100,000 different fossil species and no intermediate forms have been found, but while further discoveries are awaited, it is accepted by neoDarwinists that it really does not matter anyway as the genetic engineers have already provided enough proof that the neoDarwinian theory of evolution is the truth which they believe it to be.

Chapter 4

Bones

It may seem from the previous account of evolution that apart from there being a few gaps in our knowledge, many of the important facts and details of the past are known and understood. Unfortunately, nothing is further from the truth. Virtually all the opinions which have been expressed about past life are just that – opinions, and very little indisputable evidence has been identified so far which relates directly to the appearance of the first living cell or to any subsequent evolutionary events.

In fact it is freely acknowledged by many that the fossil evidence around which opinions have been formulated is actually quite inadequate, and in many instances there is no fossil evidence at all which relates directly to them. Instead, the reasoning which seems to have underpinned all the opinions is that because features are present in living species today, they must have evolved in the Darwinian way at some time in the past and because they evolved in the Darwinian way, it is reasonable to speculate on the evolutionary processes which must have been associated with their appearance. As the neoDarwinian theory is believed to be true and the only theory which can applied to the study of evolution, all the opinions have been developed within its framework. Some of them have even been developed beyond the stage of being opinions and are actually regarded as being proven facts. However, when such opinions are examined critically, it becomes clear that no one can make any statement at all about the nature of evolution let alone accept it as being a fact unless that statement is supported by adequate fossil evidence.

The whole essence of the neoDarwinian theory is that all life has evolved as the result of randomly-occurring changes to genes, with natural selection being the only provider of any logic which may be present in life. Therefore, if opinions, and presumably they have been logically- formulated opinions, are expressed about past life, unless there is supportive fossil evidence for those opinions, it must simply be a matter of chance if they are right or not. Yet this practice of making unsupported statements has been carried into the heart of the neoDarwinian theory itself.

One of the major tenets of the neoDarwinian theory of evolution is the understanding that life commenced as a single cell and if the theory is the truth it is believed to be, obviously that statement must be correct. But there is absolutely no direct supportive fossil evidence for this very positive statement about life. The only vaguely supportive fossil evidence for it is that which was used by Darwin when he was endeavouring to determine how life started. He found that as one retreated in time, fossils became progressively more simple and from this he deduced that if one went back sufficiently far, one would ultimately reach the most simple of all the life forms which had ever existed. If that was so, he deduced that life must therefore have started in one of those life forms. It was a logical conclusion to draw, but although the fossil evidence pointed him in this direction, his opinion on this matter was just that, an opinion, and the fossil evidence which he used to formulate it did not, and still does not, confirm it. There is no fossil evidence which shows that when life first appeared, it appeared in the form of a cell or cells nor is there any fossil evidence which shows that if life actually did appear in cells, it appeared in only one of them. Yet this belief is one of the essential planks of the neoDarwinian theory of evolution and because of this, the information which has been used to formulate it needs to be reviewed to determine its accuracy and how it came to be made.

The information which has been used to formulate the statement has come from studies of living cells of today and from them alone. Such studies have shown all the life which is present and alive on earth today is present in living cells and only in living cells. No other form of life exists on earth. Living cells are, in fact the basic units of life, and from this it has been deduced that Darwin's most simple of all life forms must have been cells. Similarly, because all living cells are also alive, it has also been deduced that those ancient cells must also have been alive. All living cells also share the same family resemblances, and from this it has been deduced that this can only be so if they have all descended from a common ancestor which, because living cells are cells, must also have been a cell. The summation of these deductions has been that only one living cell was involved in the appearance of the first spark of life on earth and that all subsequent living cells have descended from it.

These conclusions may be correct but they are completely unsupported by fossil evidence and have been formulated deductively using information which has been obtained only from the study of life which is alive on earth today.

As with the neoDarwinian beliefs about the first cell, so also with its beliefs about genes. By studying living cells, it has been found that all of today's living features are controlled by genes and that they contain all the codes of life. From this it has been deduced that all features of past life were also controlled by genes and it is not even been remotely considered that other earlier biological systems may have contained the codes and have the controllers of life in the past. It is central to the neoDarwinian theory that all evolution throughout all living time has been the result of the randomly-occuring genetic changes, yet, as with its statements about the first cell, there is no fossil evidence which actually supports this. Although genes may have been the central part of all past life, when their host cells died and disappeared, so also did they and they did not fossilise. Therefore any comments which are made about them and their past behaviour are also deductive suppositions.

Despite this, the various deductions and conclusions including those made about the family resemblances which are in cells have been generally accepted as being true, yet if all evolution has occurred by chance, how can characteristics which were present in the first cell still be present as family resemblances in all of its living descendants? If the first cell's characteristics were controlled by genes which are understood to have appeared by chance, and if chance has changed them randomly ever since, how can characteristics of the first cell still be universally present in similar forms in all currently-living cells? How can these things be? Either the family resemblances are mock resemblances which have convergently evolved to their present similar forms from a number of different primary sources of life, or they are real family resemblances which have not changed significantly since they first appeared in the first spark of life. If they are mock resemblances and are the results of convergent evolution, the neoDarwinian theory of evolution is wrong and all life has not stemmed from a single source. Alternatively, if they are real family resemblances and all living cells have descended from a single cell, the neoDarwinian theory of evolution is still in real difficulties when it attempts to explain their presence in life today. Their presence infers that some of the genes which were present in the first cell have not changed since then and within the concept of the theory this cannot be.

To establish the true status of the neoDarwinian theory therefore, each of its components need to be examined separately and their individual truths established. If one of the components is proved to be untrue it will cast serious doubts on the integrity of the theory as a whole and if more than one is proved to be untrue, the whole theory must collapse.

It has been shown that the theory has two major components. One which concerns the form of the first spark of life and the forms of the life which have descended from it, and the other which concerns the way that the spark appeared and way that subsequent life forms evolved. So far neither has been assessable within a satisfactory framework of fossil evidence and both have been formulated using information which has been obtained from today's life and today's cells. If the use of today's information to make logical comment about the past is theoretically unacceptable within the neoDarwinian theory, before the theory's components can be examined, it is first necessary to determine if it is actually proper, valid and logical to use information gathered about today's life to examine past life. If it is, proving it may seriously damage the credibility of the neoDarwinian theory.

However, if it is not valid, proving that it isn't, will strengthen the neoDarwinian theory even though it will cast doubts on the accuracy of some of its statements. It is therefore important that a clear and simple method is used to determine if it is valid to use today's information to examine the past and the results are unequivocal. It can be assessed by:

1   examining features which are present in life today

2   using the information which is obtained to make predictions about those features as they were when they were present in living individuals in the past

3   confirming the predictions by examining those features as they were in the past by examining those features in fossilised individuals.

Bones are good features to study in this way. They are easily examined in life and the accuracy of any predictions which are made about them as they were in the past, can be assessed by examining the actual bones of past individuals. Fortunately many bones of past individuals have been preserved by the processes of fossilisation and the processes of preservation have been excellent. When ancient individuals died, although their bones usually crumbled away into dust, occasionally they did not. Ground waters entered their internal spaces including the very small recesses which had previously been occupied by bone cells, and as the mud, sand or salts in the water settled out and solidified, they faithfully supported and preserved the bones including their most delicate structures. In time, interactions occurred between the ions of the calcium salts of bones and the ions of the mud or sand, and, particularly where silicon ions were involved, the original relatively fragile bones became converted into robust siliconised stone bones. If the surrounding mud and/or sand also turned to stone, those stone bones were then encased in stone and in this form they have remained faithfully preserved in almost their original condition for hundreds of millions of years. Consequently, direct and valid comparisons can be made between all modern and ancient bones.

The bones of living mammals, birds, reptiles and amphibians

However, before any comparisons can be made it is first necessary to study modern bones. It is obvious that mammals, birds, reptiles and amphibians have many different body forms and live in many different environments. Superficially, they seem to have few features which are similar in all of them but when they are examined more closely, it becomes clear that this is not so. Particularly, they have similar bony endoskeletons which classically consists of a skull and mandible, an axial vertebral column made up of a string of vertebrae which articulate with their neighbours, and pairs of fore and hind limbs which connect with the trunk through bony fore and hind limb girdles. Each limb ends in a hand or a foot and each of these have five digits. However, just as their body forms may vary quite considerably from one to another, so also do their endoskeletons. In fact most adults do not

32 EVOLUTION AND A CREATOR?

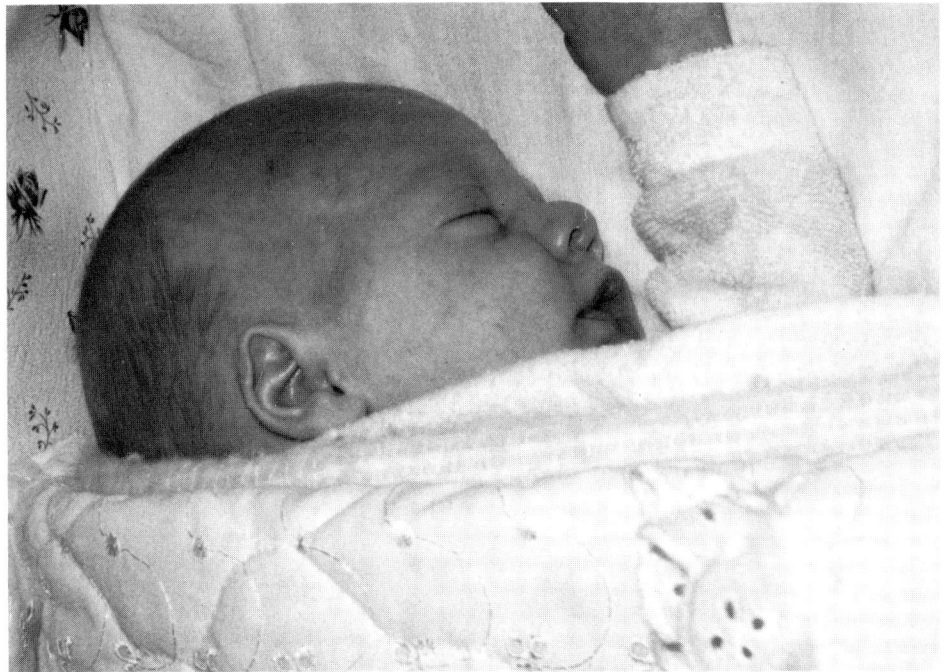

A mammal.

A reptile (tuatara).

A bird (kiwi).

An amphibian.

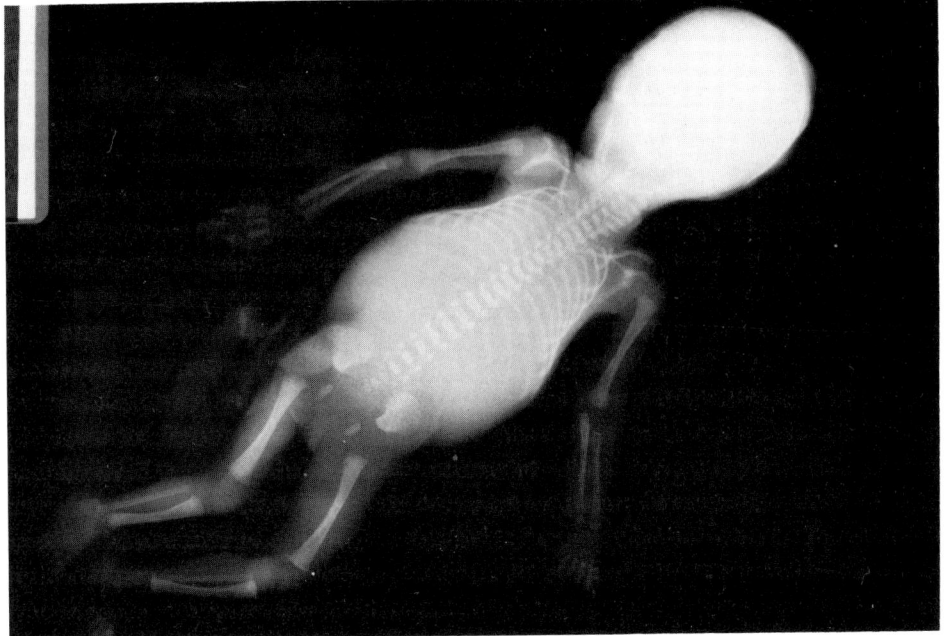

A whole body radiograph of a mammal (baby).

A whole body radiograph of a reptile (skink).

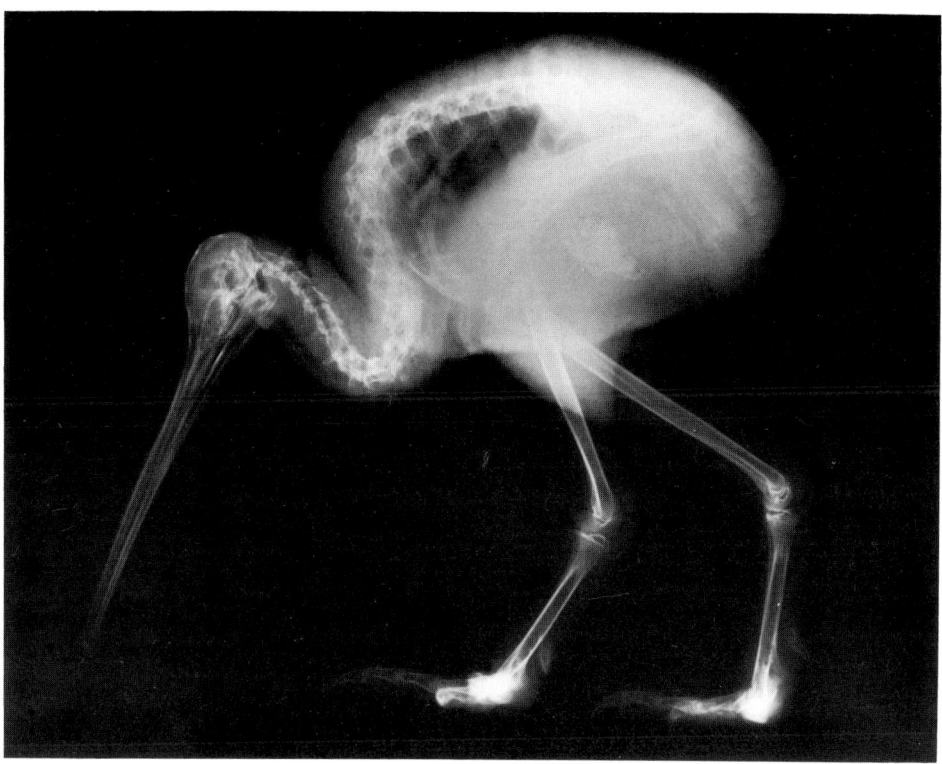

A whole body radiograph of a bird (kiwi).

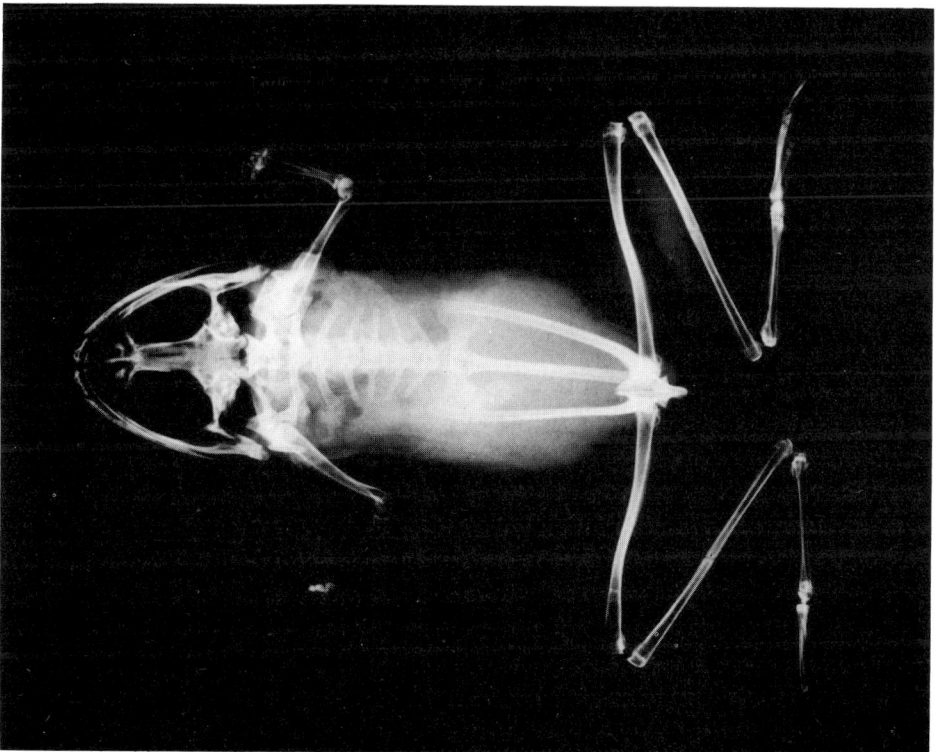

A whole body radiograph of an amphibian (green frog).

36   EVOLUTION AND A CREATOR?

seem to have this classical skeletal form at all. Importantly however, the classical form is present in all of their early embryos and it gets masked or it vanishes as the embryos grow into adults. As all early mammalian, avian reptilian and amphibian embryos have the same skeletal forms including paired fore and hind limbs with five digits on their hands and feet, collectively all mammals, birds, reptiles and amphibians, embryo and adult, are accepted as being pentadactyl tetrapodal vertebrates (PTV's) and are classified as such.

The usual way to examine PTV bones is to inspect them – hold them in one's hand and look at them, and virtually all the conclusions which have been drawn about them have been based on such examinations. However, there are other ways of examining them and when two of these are combined, a quite different perspective of bones and joints emerges. Using the technique of Xraying bones and photographically adjusting the resulting images for size, once the size differences have been focused out of the overpowering consideration which they normally present visually, it becomes clear that there are many more similar features present in the bones and joints of modern mammals, birds, reptiles and amphibians than had previously been suspected.

Similar features, however, are similar because they contain elements which are the same in each of them and elements which are different, and when the similarities which are present in comparable bones are examined and their two sets of elements are examined and compared separately,

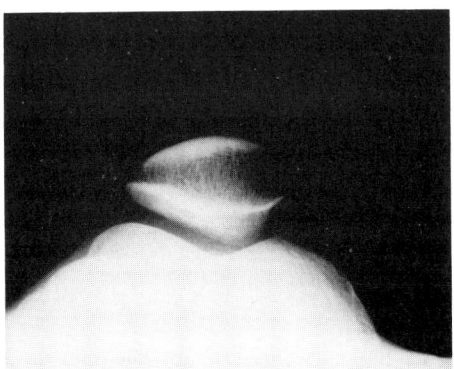 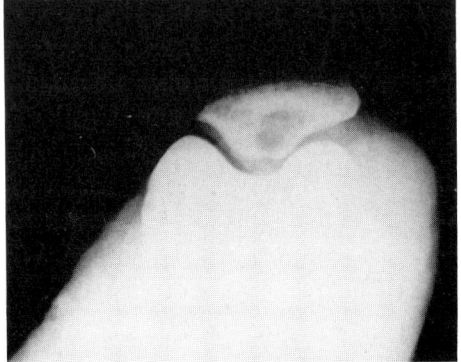

Radiographs of the knee caps (patellae) of a mammal (human) and a bird (domestic hen). When the two bones are held in one's hand and examined visually, because of the differences in their sizes and the slight differences in their shapes, it is not possible to fully appreciate how very similar they are to each other, but once they have been radiographed and the sizes of the radiographs have been appropriately adjusted, the degree of their similarity to each other becomes evident. Before they can be studied further, it is necessary to realise that within each similar feature there are elements which are the same in both of the bones and elements which are different. Each needs be identified and studied separately before it can be understand how the elements came to be the same in such different creatures and how the differences occurred. When that is understood, the real nature of evolution becomes clear.

further unexpected information becomes evident. It becomes clear that there are many elements of comparable bones which are not just similar to each other, they are exactly the same in all of them and instead of the differences occurring in a host of unpredictable ways, essentially they occur in only two ways. Either they are the result of bones fusing to one another as the embryos grow, and/or they are the result of parts or all of bones growing at different rates and reaching different sizes in adulthood.

The same findings apply to the skulls and the shoulder and pelvic girdles of mammals, birds, reptiles and amphibians though it was not possible to confirm this in the way it is possible in the limbs. As they could not be disarticulated sufficiently for the individual bones to be examined separately, they could only be examined visually in their articulated state. However, with the knowledge gained from the examinations of the limbs, it could be seen that comparable skulls and limb girdles had many components which were the same in all of them, and subject to some minor qualifications, all the differences which were found were the result of local and/or generalised growth differences and/or fusions of bones or parts of them to one another. The minor qualifications concern the temporo-mandibular joints; the inconstant number of vertebrae in PTV's, and the relatively inconstant positions of the limbs relative to specific vertebrae.

Although temporo-mandibular joints may be the same in all mammals, or the same in all birds, or the same in all reptiles or the same in all amphibians, they are not universally the same in all PTV's. With vertebrae, the number of vertebrae in individual mammals, birds, reptiles and amphibians may vary quite considerably as evidenced by the differences in the numbers of vertebrae in the spines of snakes and frogs. However in each instance the basic forms of the vertebrae themselves are constant. Slight differences may also occur in the positions of the limbs relative to specific vertebrae in the spines of different species though the limbs themselves always have the same embryonic forms in all PTV's.

It may be argued that such sweeping statements cannot be made when so few PTV species have been studied and in order to reduce the strength of this objection to a minimum, the bones of as many different mammals, birds, reptiles and amphibians as could be obtained were examined visually and radiographically by the author. These included one or more of the bones of a salamander, snake, monitor lizard, peacock, pheasant, penguin, magpie, whale and an elephant, and the bones of toads, skinks, ducks, hens, bantams, sparrows, thrushes, blackbirds, cats, dogs, mice, rats, deer, sheep, goats, cattle, horses and many many humans.

Some predictions can now be made about the forms of ancient mammalian, avian, reptilian and amphibian bones using the information which has been obtained from this study of living mammals, birds, reptiles and amphibians

If it is valid to use information obtained from the study of life today to accurately study life in the past, because there are features which are the

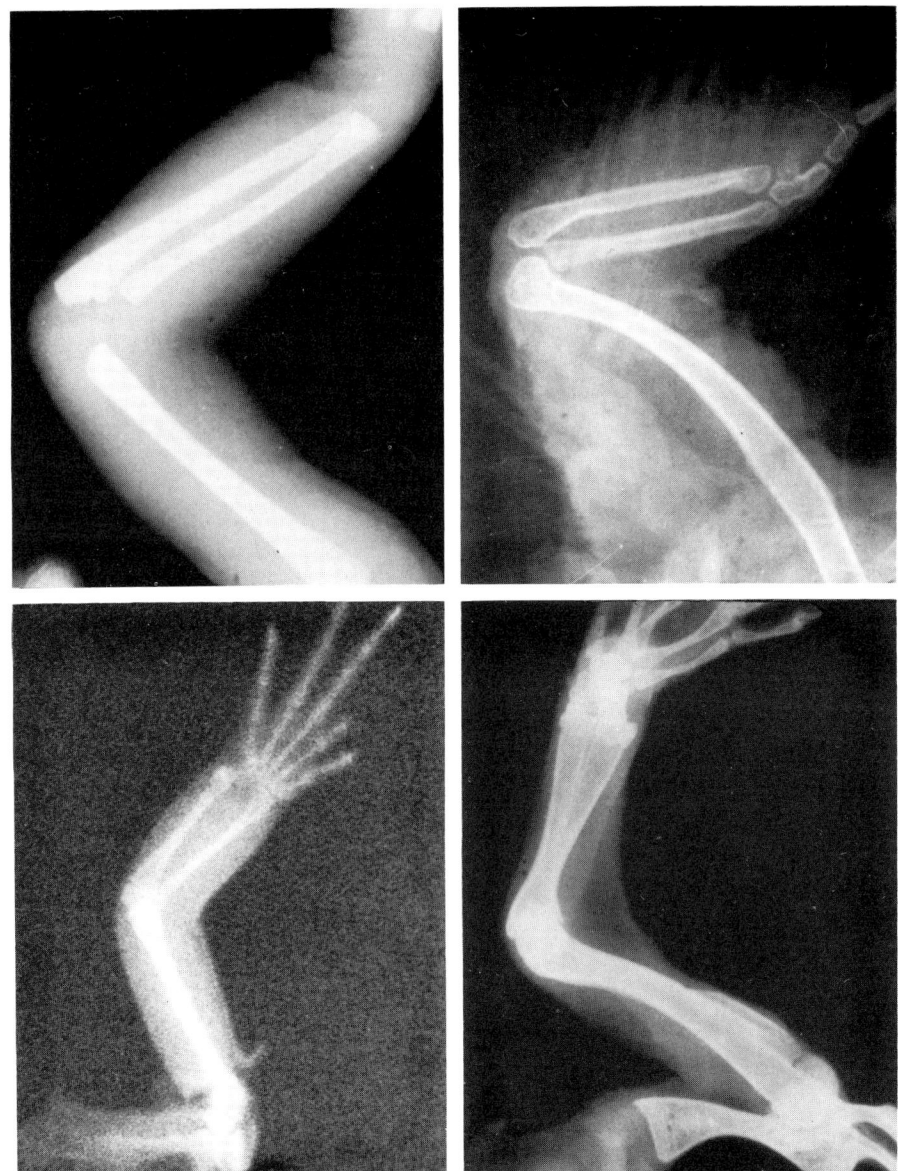

Forelimbs.
Despite the arm of the mammal (baby) (*upper left*) being relatively large, the arm of the reptile (skink) (*lower left*) being quite small, (only millimeters long), and the other limbs being somewhere in between, when they are examined radiographically and the results have been photographically modified for size, it is possibe to appreciate how similar they are to each other. Essentially, all the integral elements of all the forelimbs are the same except for the differences which are present and these occur in only two ways. Either they are generalised and/or localised differences in the sizes of the bones, and/or they are fusions of bones together which have occurred during growth.

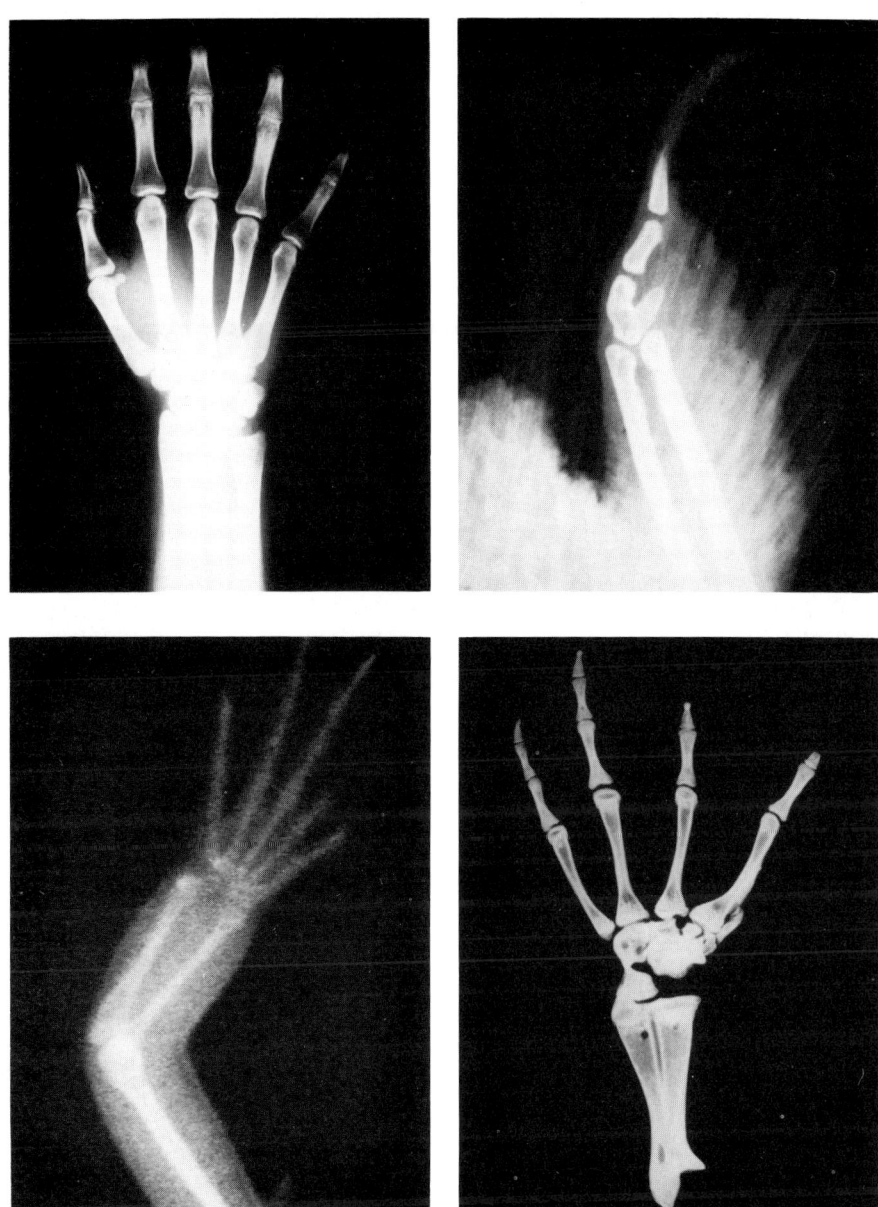

With fusions of bones there is an associated loss or a fudging of their original identity. This is seen most clearly in the bird (kiwi hands) *(upper right)*. In the early kiwi embryo hand *(not illustrated)*, five digits and a full complement of wrist bones are present, but as growth continues, a number of bones fuse together and the final result is a single-fingered hand with a reduced number of wrist bones. The single digit is tipped by a spur which in other species may be called a fingernail, a hoof or a claw.

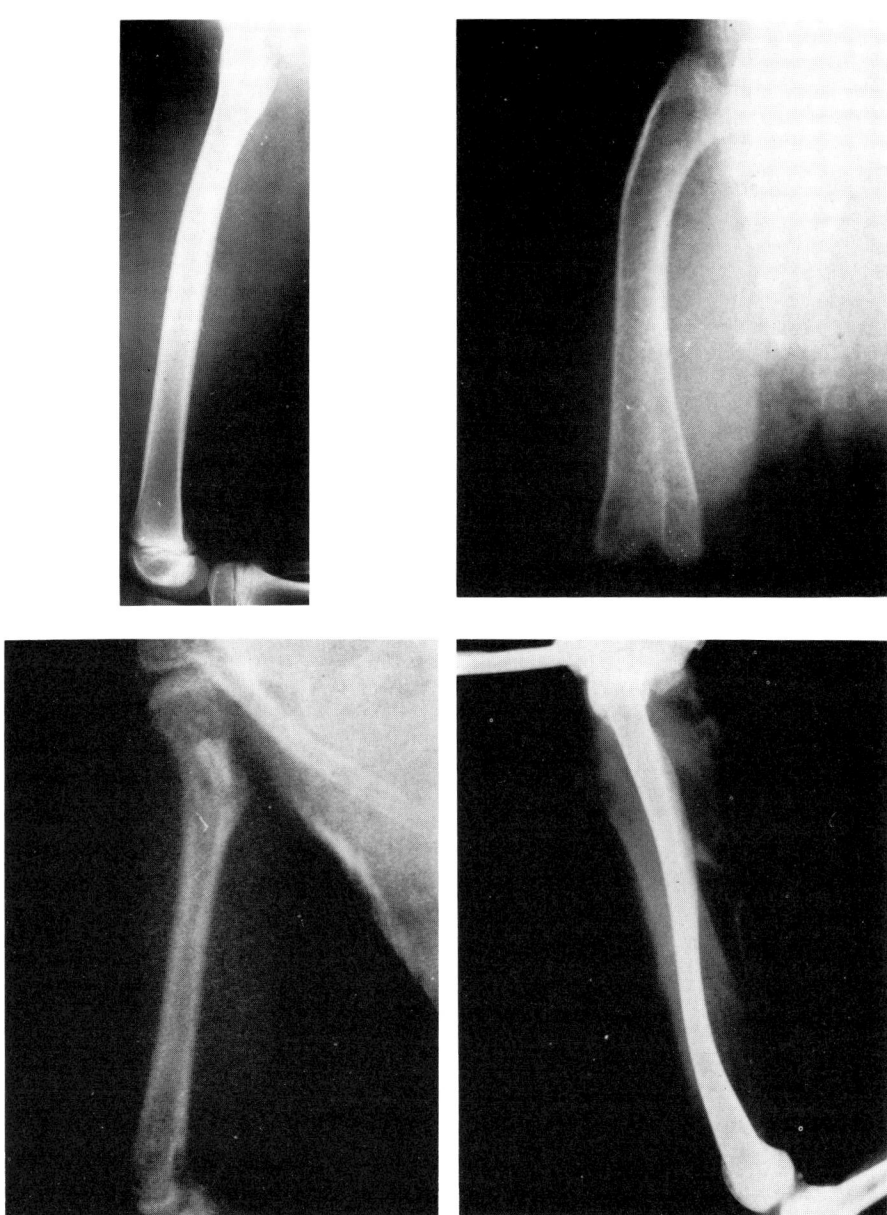

Hind limbs.
Because of the considerable size of the mammal leg (human) *(upper left)*, it could not be radiographed on a single film and the segments of the limbs of the other species have been matched accordingly. As with the fore limbs, comparable bones have many elements which are the same in all of them and the differences which are present are either general and/or localised differences in size or are the results of fusions of bones during growth with loss or fudging of their original identity.

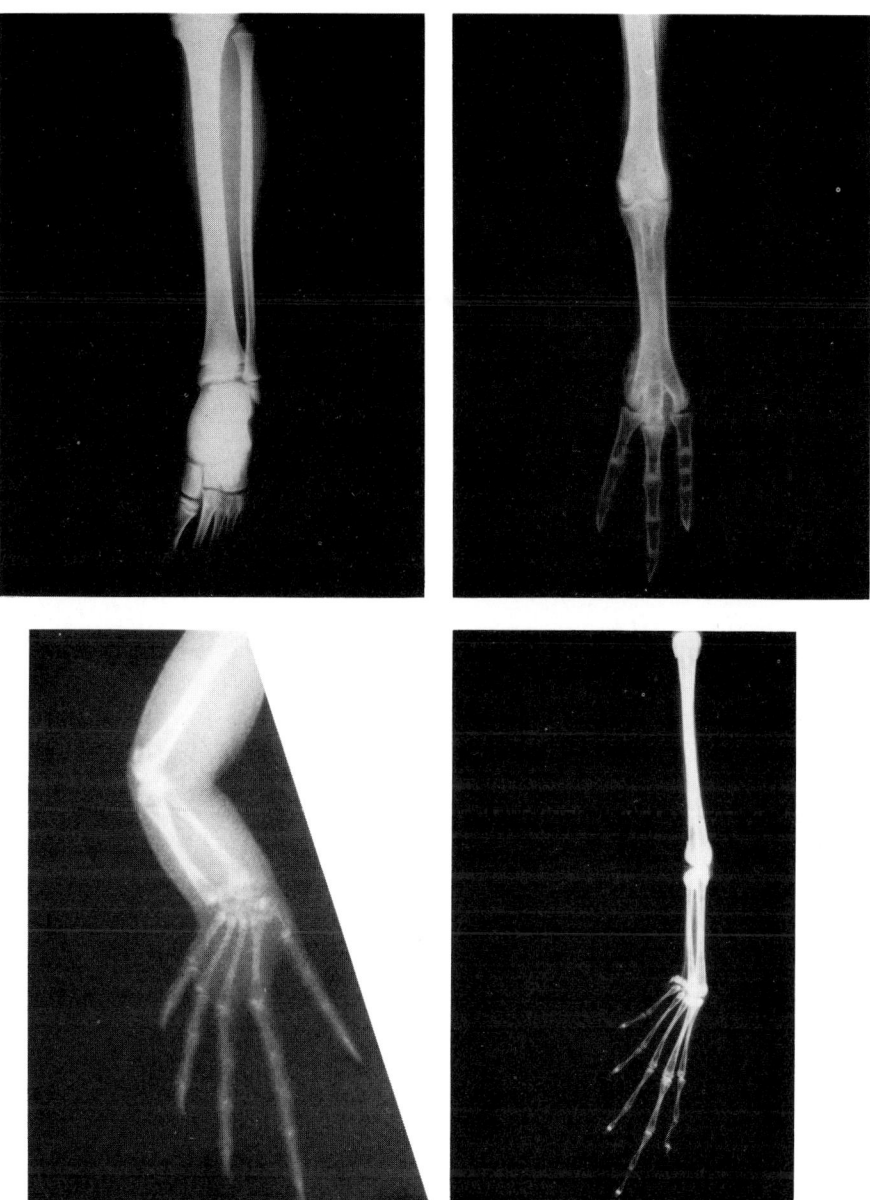

Each early PTV embryo has the same form of five-toed hind limb and this form has persisted into adulthood in three of the above species. In the adult bird (kiwi) *(upper right)* however, there are only three major toes present and these are articulating with the end of a composite bone formed during growth by the fusion of several separate bones. If enough Xrays are taken with the foot in different positions, traces of the remaining two toes and their support bones can be demonstrated. Each toe is tipped by a spur.

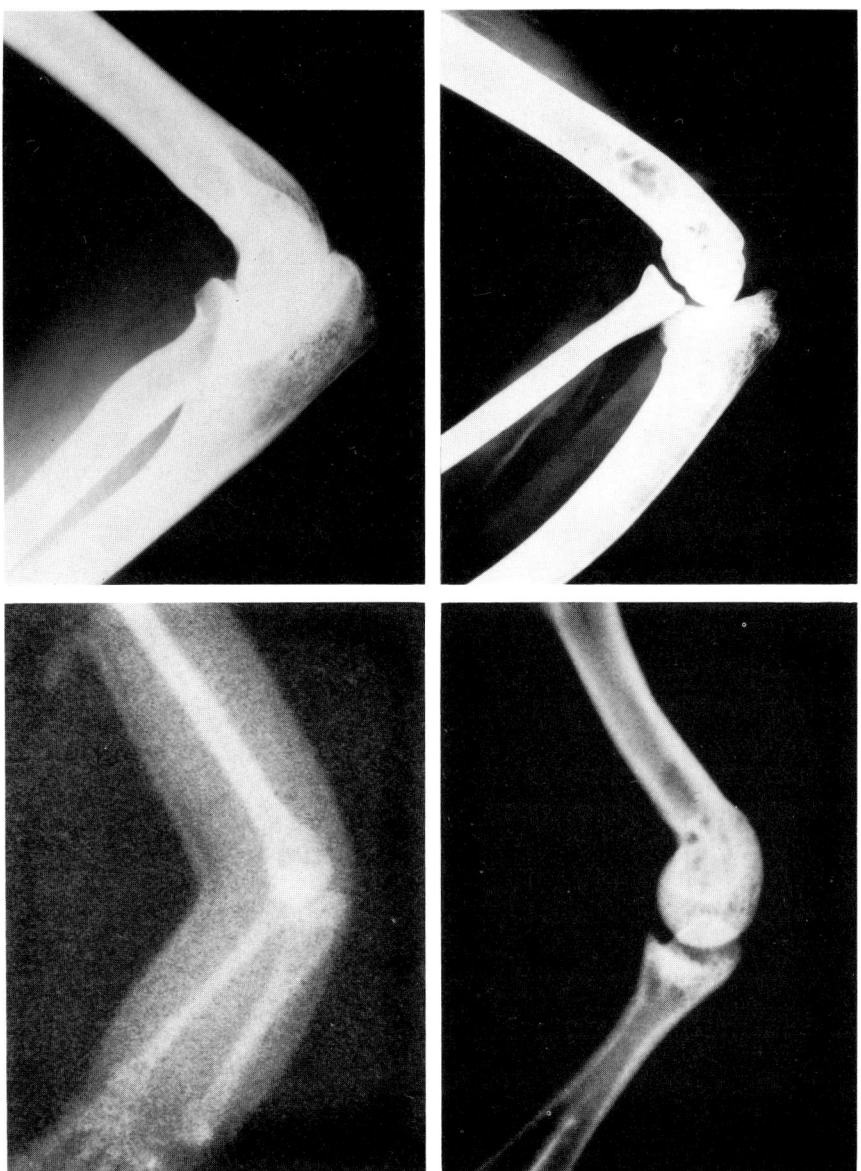

Elbows.
The elbow of the domestic hen *(upper right)* has been substituted for the kiwi's elbow. Apart from the differences in their sizes, each elbow is essentially the same as each other elbow. The most obvious difference is the fusion together of the radius and the ulna in the amphibian (frog) *(lower right)*.

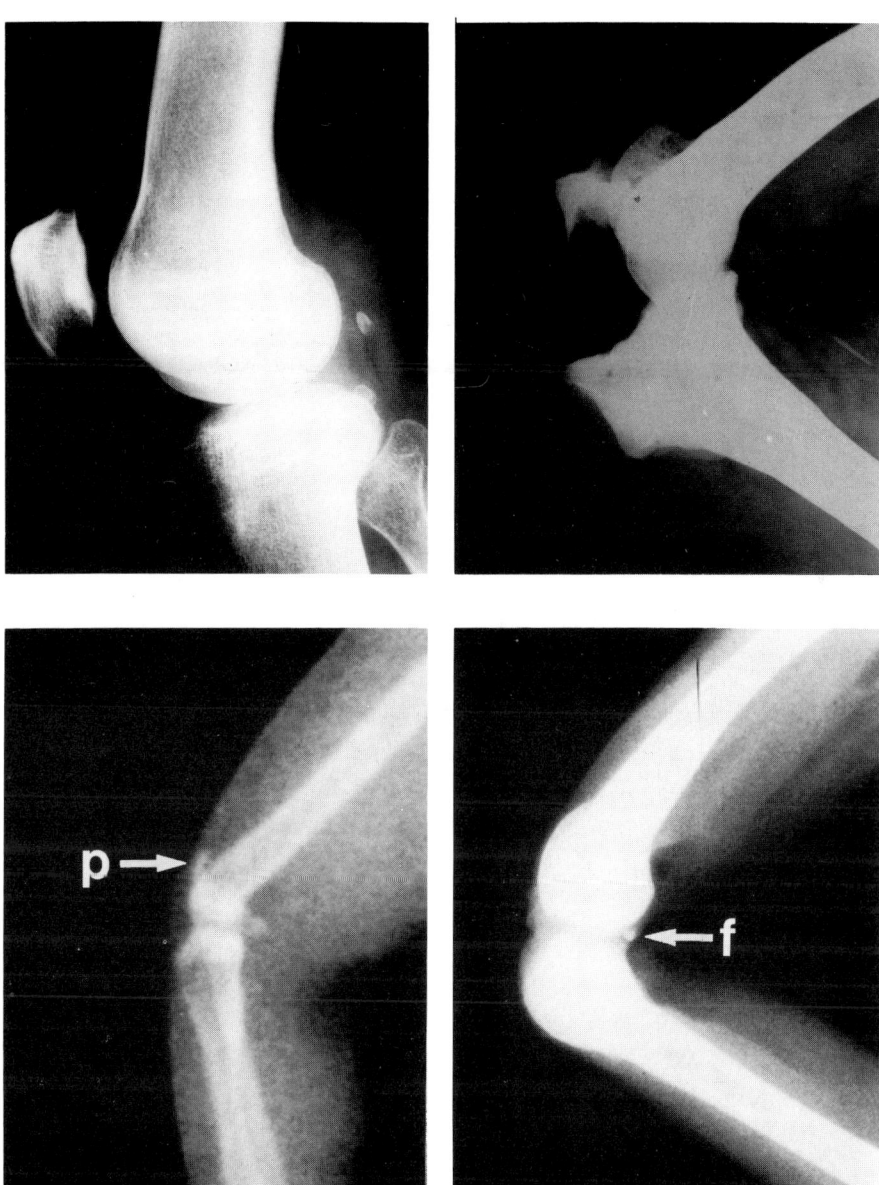

Knees
In the knees, the same findings prevail. Subject to their differences in size, each knee is essentially the same as each other knee. It is the small sesmoid bones about the knees which provide the greatest interest. All but the amphibian (frog) *(lower right)* have a patella (p) and all but the bird (hen) *(upper right)* have a fabella (f). In each species the same bones are located in the same tendons and muscles and this reflects the fact that not only are comparable bones and joints of mammals, birds, reptiles and amphibians effectively identical, the soft tissues which surround and move them are also the same – subject always to the differences which are present. From the overall appearance of the knees it is reasonable to predict that when enough birds and amphibians are radiographed, some at least will be found which have patellae and fabellae.

44 EVOLUTION AND A CREATOR?

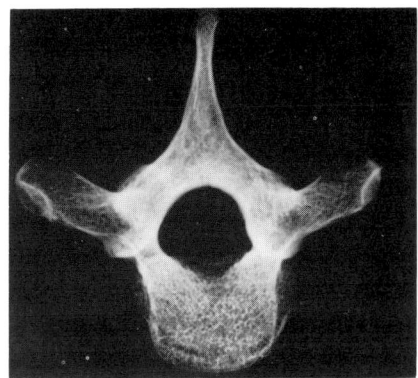
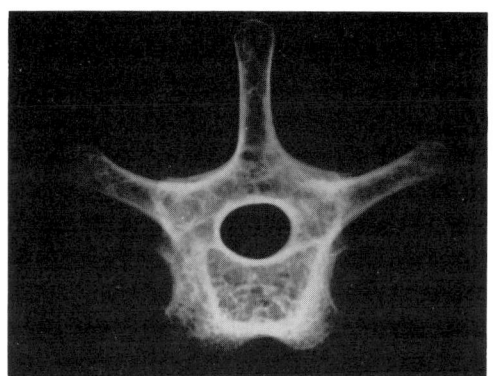
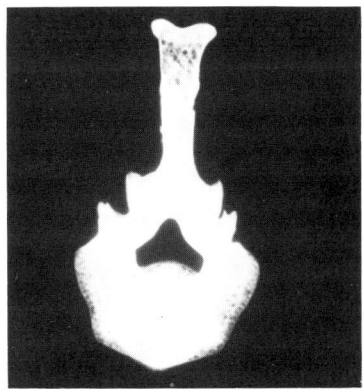
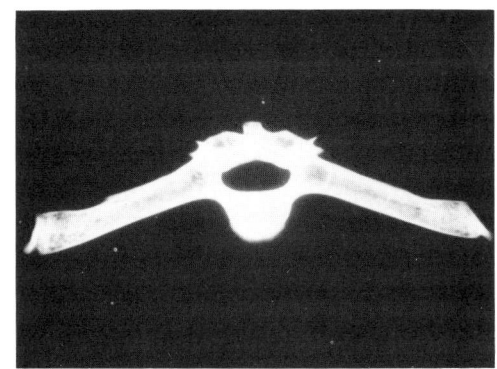

Vertebrae.
In this instance, the vertebrae which have been radiographed are human, ostrich, tuatara and frog vertebrae and the bones follow the same rules as the limb bones. Comparable vertebrae have the same basic patterns and all the differences which are present are the result of localised and /or generalised differences in the size of parts or all of the bones and/or fusions of aspects of bones to one another which have occurred during growth. In the mammalian (*upper left*) and avian (*lower left*) vertebrae, the only easily observable difference is the difference in the diameters of the canals of the nervous system. In the reptile (*upper right*) and the frog, the spinous processes are larger and smaller respectively than their counterparts in the mammalian and avian vertebrae and the transverse processes are small or are masked by them having fused with adjacent ribs.

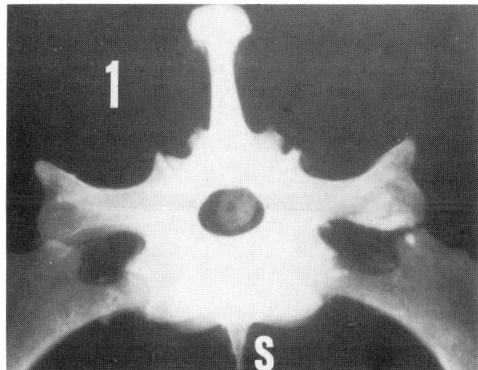
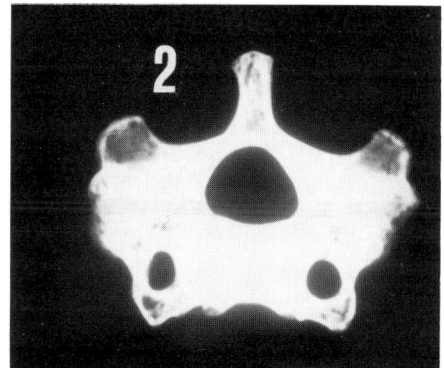
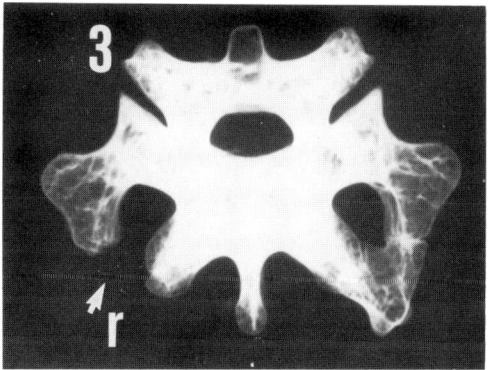

Other selected vertebrae

These show how fusions and localised differences in the growths of comparable parts of vertebrae change their appearances without changing their basic forms. The first vertebra has ribs and it can be seen that the spinous process and the transverse processes have slightly bulbous ends relative to the same structures in the vertebrae shown opposite. There is also a spine (s) projecting ventrally from the centrum. This is an inconstant feature in vertebrae and is a genetically-controlled response to the demands of muscles and tendons which are attached in the area rather than being as absolutely fundamental to vertebrae as are the transverse and dorsal processes.

In the second vertebra, the ribs have fused at their attachment points and the balance of the ribs have vanished. The shape of the spinal canal has been altered by growth variations as has been other localised parts of the vertebra.

In the third vertebra, more growth variations have occurred in some locations quite noticably changing the form of the transverse processes for example without affecting their basic nature. One diminutive rib (r) has been removed to demonstrate its form.

46   EVOLUTION AND A CREATOR?

same in all modern PTV bones and joints, it can be predicted:

1   that all living PTV's have the same features because they have inherited them from a common ancestor

2   that the features were present in that ancestor in the same forms that they are present in its living descendants today.

3   that all the generations of mammals, birds, reptiles and amphibians which have been intermediate between the common ancestor and its living descendants also had the same features in the same forms that they have now.

4   that convergent evolution has not played any part in the continuing inheritance of these features by the intermediate generations.

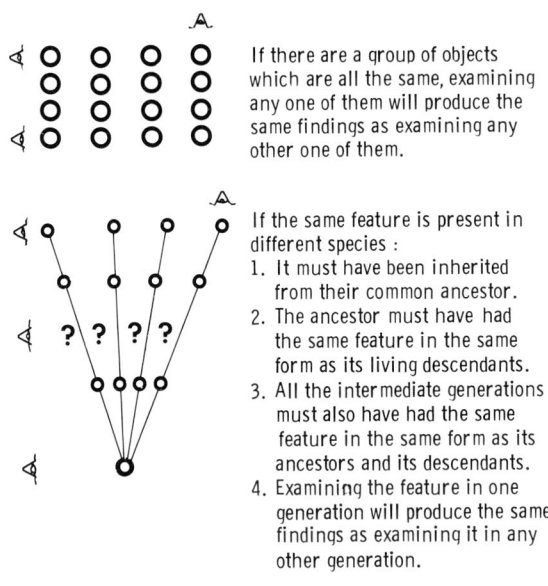

The triangle of study

Assessing the accuracy of these predictions

Within the neoDarwinian theory of evolution, it is accepted that different species have similar features either because the features have been convergently evolved that way by unrelated species, or they have been inherited from a common ancestor. The presence of so many similar features in the comparable bones and joints of mammals, birds, reptiles and amphibians must eliminate convergent evolution as being a reasonable explanation for their presence. Therefore, the features must have been inherited from a common ancestor, and in the case of PTV's, it is generally believed that that ancestor has been found and identified. Three fossilised amphibian species of the Devonian geological time of 425-375Myrs ago have been discovered, each with a bony endoskeleton similar to that of living PTV's, and it is accepted that one of these species was almost certainly the ancestral species of all subsequent PTV species. It is also accepted that if none of them was the actual ancestral species itself, each was very closely related to it and that it had the same bony endoskeleton the ancestral species had. So far however, none of the skeletons of the three fossilised species have been examined using the information which has been obtained from this radiographic-photographic study. Therefore the absolute accuracy of any predictions which are made about it cannot therefore be completely confirmed. However, the fossilised skeletons of the three Devonian PTV's have features which are similar to those which are universally present in all living PTV's today, and it is their presence which has been used to deduce that one of them was the ancestral species of all subsequent PTV's. Therefore, if the concept of similarities is correct, that is, similar features contain elements which are the same in all of them and elements which are different, it is certain that within the similarities of the endoskeleton of the ancestral species, there are elements which are identical to those which are present in its living descendants. If this were not so, the features could not be similar to the features of the present in the ways that they are. Deductively therefore, it would seem that the first part of the predictions is correct. It is possible to use living information to predict some aspects of the past, and details about features of the common ancestor can be determined by examining and determining those similarities which are present in its living descendants and determining the elements of them which are identical in all of them. These elements must be present in the skeletons of the common ancestral species.

Predictions have also been made about the nature of the features which were present in the bones of all the generations of PTV's which have been in the direct intermediate lineages between the Devonian common ancestor and its living descendants. A visual study of the fossilised bones of approximately thirty individuals from those many generations, showed that in all the various bones which were examined, the same features were present as those which are present in the comparable bones of living PTV's. It was also found that all the differences which could be identified in the fossil bones also occurred in the same ways that they do now (by fusions and/or by variations of growth). Though only a small number of fossilised bones were examined, from the consistency of the findings it seems reasonable to presume that when other discovered fossilised PTV bones are examined, the same will be found for them and if this is so, it is awkward for the neoDarwinian theory. If life has evolved as the result of random genetic changes, proving that the predictions which can and have been made about ancient bones from examinations of their descendants are valid for all intermediate generations of PTV's, not just the discovered ones, will be of the greatest importance to the theory. It will never be possible to absolutely prove such predictions as fossil remains of all the generations of past PTV's simply do not exist. However every time a new PTV fossil is discovered which has all the features which are identical to those present in all the already discovered

PTV fossils and which also lacks any intermediate forms of bones of the kind Darwin needed to have when he was trying to prove his theory, it is a mark against the neoDarwinian theory of evolution and a mark for using present life studies to study ancestors. Yet from the Xray studies and the observations which have been made so far, it seems reasonable to expect that if the predictions made about the intermediate generations are correct, every future PTV discovery will be found to have bony features which are similar to those of existing discoveries and within those similarities, there will be features which are exactly the same as those which are present in bones today. It will also be found that the differences which exist between the similar features have all occurred in the same ways that they do now. The skeletons and the individual bones of the new discoveries will not have unpredictable forms as some should have had they evolved in the creeping neoDarwinian way. If they had evolved in this way, at least some of them should have bony forms which were quite different to the known PTV skeleton and if any of these new forms are ever found, it will prove the neoDarwinian theory. But, every time another new extinct PTV species is found which has the same predictable skeletal form as all the other discovered PTV's, it will continue to confirm that for the PTV skeleton at least, creeping neoDarwinian evolution has not happened. Evolution has certainly occurred. There is no question of that, but within the PTV group of species, their endoskeletons have not changed since the first PTV skeleton came into being in the Devonian ancestor, subject only to the minor differences which have occurred since then, and in each instance, these have occurred in the same ways as those which can be identified in PTV skeletons today. Against neoDarwinian expectations, the PTV skeleton has not appreciated since the Devonian. Instead, the only changes which have occurred to it have been relatively minor modifications or depreciations of its form.

The study of the bones of living PTV and the bones of their ancestors was undertaken to determine if it is valid to examine today's living species and use the information obtained to accurately study past life. As much as it can be proved within the confines of existing fossil discoveries, it has been found to be valid for PTV bones and skeletons and there is some deductive evidence which supports this.

All living PTV's have their present skeletal and individual bony forms because they have inherited them in their present forms from their parents who must also have had them in their present forms. For the parents to have them in the same forms as their offspring, they must have inherited them in the same forms from their parents, who in turn inherited them from their parents, and so on and so on without change back through converging lineages of antecendent mammals, birds, reptiles and amphibians to their ancient common ancestor. It also had to have had the same skeletal and individual bony features as its living descendants otherwise they could not all have the same features with the same forms that they do have. For the same reason, all the generations of PTV's intermediate between the common ancestor and today's species also had to have had the same skeletal and bony features otherwise some of their living descendants would have forms which are different to that which we know they do have. The universal presence of the same bony features in all living PTV's; the presence of the same features in the common ancestor; the universal presence of the same bony features in all the PTV bony fossils which have been discovered and studied so far, and the probability that the same features are present in all undiscovered PTV fossil species, makes it extremely unlikely that any creeping evolution has been involved in any way in the continual inheritance of the same PTV skeleton by so many different species for so long. The only possible group of PTV's which could have skeletal forms which are different to that which we know is a lineage or lineages of descendants of the ancient common Devonian ancestor which have diverged from the known PTV lineages and which have become completely extinct and have vanished without trace. That hope for the neoDarwinian theory of evolution still exists in the undiscovered segment of the fossil records though as further evidence is presented, that hope becomes smaller.

There is some other deductive evidence which supports this view about PTV skeletons. Using their own form of logic, Creationist taxonomists from Linnaeus onwards, established that there were common patterns in the skeletons and individual bony features of living mammals, birds, reptiles and amphibians. The names which they gave comparable bones in quite different species reflect this. All had a humerus, or a femur or a mandible etc (unless as in snakes for example, the bones had failed to grow after they had appeared in the early embryos), and when they were classifying new species, they continually sought and found these expected patterns and were never disappointed. Similarly, post-Creationist taxonomists, using neoDarwinian logic, if such a thing can exist in a randomly-changing system, sought and still seek exactly the same patterns in any newly discovered PTV's and they also have never been disappointed. What neither group discerned was that the differences which are present in PTV skeletons and bones and which mask the many elements which are the same in them, have always occurred in the same limited number of ways. Great diversity of form is not a feature of PTV skeletons.

Conclusions

The skeletons and the individual bones of living and extinct mammals, birds, reptiles and amphibians were studied in an endeavour to establish if the neoDarwinian theory of evolution is flawed. Both major statements of the theory have been fashioned around a combination of a limited amount of fossil information and a lot of information which is obtained from life today. If the neoDarwinian theory is the proven truth which it currently is believed to be, one cannot use living information to make coherent comments about the past. However as has shown, it is valid to accurately predict details of the skeletons and the bones of ancestral PTV's by examining the bones of their living descendants.

This is an embarrassment to the neoDarwinian theory but because the individual components of the theory have not been examined and their individual truths determined, it certainly does not prove that the neoDarwinian theory is wrong. However it does show that the creeping form of evolution which is an essential part of the theory has never involved the PTV skeleton. It also shows that because there are identical features in the bones and skeletons of living PTV's, the controlling genes of these features must also be the same in all living PTV's, and as the features of the bones of the past were identical to those of modern bones, those past features must also have been controlled by genes rather than by any other biological system. Those genes must also have been the same then as they are now and none of them have been ravaged by randomly-occurring changes. Instead, they have been inherited in their same forms by many individuals in many generations of many different species for many many years and they have proved themselves to be very stable genes indeed. In fact, the only ones which have ever changed since they first appeared in the common Devonian PTV ancestor have been the ones which have formulated the bony species differences which have appeared and they do not appear to have changed randomly. Instead, they seem to have changed predictably as the narrow range of species differences options shows. It is only the amount of their change which has been unpredictable and even then, the changes have all occurred within definable limits. The neoDarwinian theory also has great difficulties in explaining this.

In the pursuit of the truth about the neoDarwinian theory, what has not yet been proved is whether it is valid to examine any features of any biological systems of related living species, not just bones, identify the elements which are identical in them and use them to accurately study the same features in their ancestors. Although much more proof is needed before the technique is acknowledged to be valid for all similar features in all biological systems of all related species in all kingdoms of life, the necessary proof for this is available and some of it becomes apparent when bone, the tissue which is present in the body in the form of bones is studied.

Chapter 5

Bone

Bone has two quite different functions in the body. It is the tissue which forms rigid bones and it is also a very complex dynamic biochemical system. Because the proper functioning of the biochemical aspect of bone is essential to the overall health of the individual, bone has been fairly widely investigated in mammals, birds, reptiles and amphibians (PTV's) and it has been intensively investigated in humans.

Bone is always studied when bones are Xrayed. Allowing for some variations in the concentration of bone tissue in bones, all living PTV bone is radiographically the same (see the various Xray's of bones) and this implies that bone, the biochemical system, is also the same in all living PTV's. Subject to the presence of some species differences, this supposition has been found to be true. Apart from the presence of some slight differences in some species, the embryology, growth and development, physiology, biochemistry and the hormonal and metabolic features of bone including its interactions with the other biological systems of the body, are the same in all living PTV's.

As bone was studied to determine if it is feasible to study life of the past by studying present life and projecting the information into the past using the above information, it is now possible to make some suggestions about bone as it was in the past. The accuracy or otherwise of the suggestions can be evaluated by examining fossilised bone.

Much of the information which is available about living bone is actually functional or dynamic information and because bone's activities interlock with so many of the other dynamic biological systems of the body, if it can be proved that the predictions which can be made about ancient bone are correct, it substantially increases the probability that it is valid to examine any biological system of earlier PTV's by examining the same system in their living descendants and applying the information back to their ancestors.

There are many features of bone which are identical in all living PTV's, including their interactions with other biological systems of the body. As in the previous chapter, using this information it can be predicted:

1   that all the identical features have been inherited from the same ancestor

52 EVOLUTION AND A CREATOR?

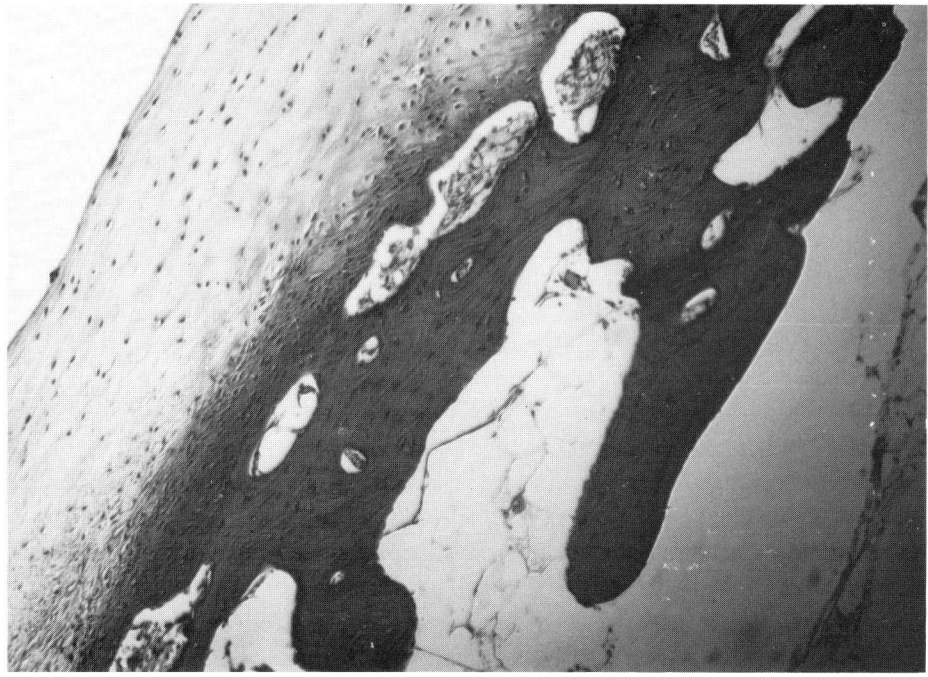

Bone.
A histological section of human bone. If the embryology, growth and development, biochemistry, physiology and hormonal control of bone and its interactions with all the other biological systems of the body are examined element by element, some minor species differences will be found in each of them, but allowing for these, all mammalian, avian, reptilian and amphibian bone is the same. This can only be so if the genes which control the elements are the same. (magnification x 400)

2   that the features were present in that ancestor in exactly the same forms that they have now

3   that all the generations of mammals, birds, reptiles and amphibians which have been intermediate between the ancestor and its living descendants also had the same bone including the same interactions with the other biological systems of the body, subject only to any species differences which may have been present.

4   that convergent evolution has not been involved in the continual inheritance of the same elements of bone by the descendants of the PTV common ancestor.

To prove that these predictions are correct, it is now necessary to study the bone of the Devonian common ancestor and the bone of the generations of PTV's which have been intermediate between it and its living descendants.

As has been shown for bones, fossilised bone is the actual tissue of the ancient individuals changed only by the processes of fossilisation and it can be examined as such. However because of its stony nature, it does not yield its secrets as easily as bones. Despite this, the consensus of the investigators who have examined fossilised bone is that all the fossilised PTV bone which has been found so far is the same as all other fossilised bone. This includes the biochemistry of its complex calcium salts, its embryology, the cells which formed it and as many of its other elements as they have been able to assess.

Deductive reasoning supports this finding. Subject only to the few species differences which can be found, today's bone including its interactions with the other biological systems of the body is the same in all living PTV's. It is the same in them because it has been inherited by them in that way from their parents who themselves must have had their bone in the same form as their offspring have. In turn the parents inherited the same bone from their immediate forebears who in turn inherited the same bone from theirs and so on and so on back through converging lineages of antecendants to their common ancestor, the amphibian of the Devonian. It also must have had the same bone as its living descendants otherwise they could not have the bone which they have. For the same reason, all the generations of PTV's which have been intermediate between the Devonian ancestor and their living descendants must also have had the same bone and the same interactions with the other biological systems of their bodies otherwise some of their living descendants would now have different bone.

The genes of bone must also be the same in all PTV's, subject to the few which have changed and have formulated the species differences, and if bone has not changed since the Devonian, its genes must also have been the same in the past as they are now. Their lack of change over such a long period of time has proved that they are very stable genes, certainly as stable as the genes of PTV bones, but studies of other fossilised bone show that they are even more stable than that. It has been found that regardless of where the fossilised bone has come from; regardless of whether it is fossilised fish or PTV bone and regardless of when the bony individual lived, all known fossil bone of all geological times is the same. No known fossilised bone is any different to any other known fossilised bone and there has been no apparent change in bone (or by inference its genes) since the first bony fish fossilised 600Myrs ago. The genes of bones may be stable, but the genes of bone are even more so.

This finding brings forward a slightly different form of deductive reasoning to that which has been used so far. Previous deductive reasoning has taken living information and projected it into the past, but the reverse is now possible. Because all fossil fish and fossil PTV bone is the same, and because all living PTV bone is also the same (subject to any species differences which may be present), as all living bony fish are descendants of ancient bony fish, if bone has not changed with the passage of time, it can be deduced that all living fish bone is the same as all living PTV bone (again, subject only to the species differences which are present). Studies, including radiographic studies, appear to confirm this. Others have shown that, subject only to the presence of some species differences, the embryology, growth and development, biochemistry, physiology, histology and other elements of living fish bone are the same in all bony fish and are the same as in all living PTV's. Radiographic studies by the author of the living bone of eight species of fish shows that bone was the same in all of them subject to minor variations in its concentration.

The species differences which exist between fish and PTV bone are slightly more evident than those which exist between different PTV species (and presumably exist between different fish species), and they have not been analysed to the same degree of intensity as those of bones, but it does appear that they follow the same general pattern. Each individual element of bone has changed away from its universal form in only a very limited number of ways though these ways could not be categorised in the same way that it has been possible for bones. Within each element however, no evidence has been found nor presumbly will it be found, that the genes of the species differences have ever changed in a random manner. As each individual element of bone is analysed and its species differences are assessed, it is probable that the differences will all be found to have ocurred in a few definable ways within predictable limits.

Bony fish.
A skeletal form is different to that of mammals, birds, reptiles and amphibians but, subject to some minor differences, differences which are principally differences in size and the ways that bones fuse to one another during growth, most, if not all fish skeletons appear to be the same as each other. The radiograph also shows that radiographically, fish bone is the same as mammalian, avian, reptilian and amphibian bone.

Other PTV biological systems

The findings for bones and bone also seem to prevail in other PTV biological systems. When biological systems of living PTV's are examined, system by system, unit by unit, and their species differences are identified and isolated, it becomes clear that the majority of the elements of each system or each unit of each system, be they of anatomical, physiological, biochemical, metabolic or hormonal origin, are the same in mammals, birds, reptiles and amphibians. For example, the histology of the various forms of smooth or involuntary muscle, striated or voluntary muscle and cardiac muscle, are the same and the species differences which are present are few. The arteries, arterioles, capillaries and veins and the endothelial cells which line them are the same, allowing for the differences in their distribution and their differences in size in different species. Nerves and their neurochemicals are the same in many parts of the central and peripheral nervous systems of PTV's though the differences in the sizes of the brains and the ways that some species use them, tend to mask this. Many of the structures associated with bones and their joints are the same including the forms and the arrangements of the muscles which move the bones, the ligaments which surround the joints and the structures which are inside the joints. However when studying a system such as the voluntary muscle system, because different workers have never fully harmonised their studies, the names used for the same things in different species sometimes obscure this in a confusing way. An illustrative example of this occurs in bones. In veterinary surgery, the femur of a deer or horse is also called the stifle and it requires more than a passing knowledge of biology to be able to comprehend that in instances such as this, quite different names describe exactly the same feature in different species.

Superficially, some biological PTV systems, however, do not seem to follow the general pattern and everything seems to be different in the system with few similarities being present, but, as was found in the Xrays of the whole bodies of mammals, birds, reptiles and amphibians, when detailed examinations are made, they reveal that this is not so. There are many components present which are the same in all of them. The differences, particularly size differences simply mask them. For example, the skins of living mammals, birds, reptiles and amphibians appear to vary considerably from species to species, but once the differences have been identified and isolated, the balance of the features are found to be the same. The cells which form the skins are the same as are the skin glands, the muscles associated with them, the nerves which innervate them and a number of other elements.

Convergent evolution cannot explain this. All the features which are the same in all living PTV's must have been present in their common Devonian ancestor in the same forms that they have now and have been inherited from it. They also must have been present in the same forms in all of its descendants which have been intermediate between it and its living relatives today otherwise today's relatives could not all have the same elements in the way that they do. This is despite the fact that the common ancestor lived approximately 400Myrs ago and millions of generations of direct descendants have lived and propagated since then.

The findings in fish follow the same pattern as the findings for PTV's. The immediate antecedants of the Devonian ancestor were fish and their descendants are alive in the seas today. When their biological systems are examined, system by system and the elements which are the same in those elements are found, the triangle of study permits it to be determined that many of these elements were present in ancient fish in the same forms that they have now and that they were passed across unchanged to the ancestral PTV in the same way that elements of bone were. The triangle of study also permits it to be determined how many of the elements of the ancestral fish features changed at the time of their transmission into the ancestral PTV then subsequently remained stable in its descendants. An example of these changes and stabilities can be found in the hormone of bone.

Fish use the pituitary hormone prolactin and the secretions of the organs of Stannius to control their bone, while PTV's use parathyroid hormone (assisted in amphibians by prolactin) for the same purpose. During the transference of the genes of fish into the ancestral PTV, the genes of the organs of Stannius and their secretions vanished and in their place appeared the genes of parathyroid glands and their secretions. Prolactin however remained common to subsequent fish and subsequent PTV's and has persisted unchanged ever since. As a result it is now common in all fish and all PTV's today. Therefore while some hormones and their associated genes disappeared or appeared in the ancestral PTV's, the genes of the hormone prolactin which acted in concert with both the old and the new hormones remained unchanged.

Conclusions

Bone (and other tissues) have been studied to provide further evidence that the neoDarwinian theory is flawed, probably fatally so, and this evidence has been provided. As with the genes of bones, the genes of bone have not been involved in any form of creeping evolution and have exhibited extreme stability for over 600Myrs even while the forms of the bones changed from fish to PTV forms. Because of this proven genetic stability, despite the technique being impossible within the concept of the neoDarwinian theory, it is practical to study living bone and its interactions with other biological systems of the bodies of living vertebrates and project the information into ancestors.

Within the concept of the neoDarwinian theory it is also impossible to use past biological information to predict the present, but where genes have never changed since they first appeared in life, the studies of fossil bones and bone shows that this can be done.

These findings cast a huge shadow over the neoDarwinian theory of the randomness of genetic changes, but however large that shadow may be, the theory is too entrenched in Science for it to be destroyed by the evidence against it which has been presented so far. Much more evidence of its inadequacies is needed before the theory is destroyed. That evidence can be found in cells.

Chapter 6

Cells

It is one of the central tenets of the neoDarwinian theory that the first cell of life and the genes within it appeared by chance and that all subsequent evolutionary changes have also occurred following chance changes to genes. Because of this, despite the proven stability of the genes of bones and bone, within the neoDarwinian theory, it is completely unacceptable to acknowledge that those genes are permanent genes. They may be extremely stable but within the theory they cannot permanent as permanent genes cannot exist. The only genes which could even remotely be considered to be permanent, would be genes which were present in the first cell of life and are still present in life today in the same forms which they had then. If such genes are present in life today, they have to be reside in living cells and they have to reside in all living cells unless in some instances they have disappeared with the passage of time. However even their disappearance is impossible if they are permanent as by inference if they are permanent, they can never disappear. Therefore if they exist, they have to be universally present in all living cells. Living cells therefore now need to be studied.

All living cells are small membrane-enclosed compartments which contain concentrated solutions of chemicals. They are the smallest units of life on earth and examinations of them have shown:

1   that they all have the same cell membrane. Each is constructed of the same phospholipids and proteins which are always arranged in the same distinctive three-layered way. Each cell membrane forms the complete outer boundary of its cell and is biologically active. In some cells, particularly plant cells, additional biological material may be layered on to the outside of the cell membrane but any such material is additional to the cell membrane and does not replace it. A membrane similar to the cell membrane may form when some phospholipids are mixed in water, but whatever shape they may form, they lack biological activity and biological activity is as essential to the cell membrane as are all its other components and properties.

2   that all cells produce energy in the same form and in the same way. Each cell produces biologically-available energy in the form of adenosine triphosphate – ATP by breaking down glucose by the process of anaerobic

glycolysis. ATP is also formed in many cells by aerobic gycolysis, but unlike the processes of anaerobic glycolysis, aerobic glycolysis is not universally present in all living cells.

3   that all cells use ATP to maintain their healthy living state. The basic ATP-using mechanism is a protein-synthesising facility called a ribosome and the activities of ribosomes are controlled by RNA – ribonucleic acid. As new proteins are required by the cell for its processes of living, they are manufactured by the ribosomes and it seems that all cells use the same proteins for their basic processes of living and it is the presence of additional proteins which gives each cell its individual characteristics.

4   that all cells have a genetic mechanism of DNA -deoxyribonucleic acid, which controls the life of the cell including the initation and the completion of its propagation into its immediate descendent generation of cells. Many cells die before this part of their lives happens, but in cells which do propagate themselves, their genetic mechanism controls all aspects of this process including its initation, and as each cell reproduces itself, the cell's genetic mechanism transfers itself and all of its associated biological processes into the next generation of living cells.

The studies of bones and bone have suggested that it is valid to study past life by examining present life and projecting the information into the past. Therefore because all living cells have the same four identical biological systems, it can be predicted:

1   that they have all inherited them from the same ancestor
2   that they were present in that ancestor in exactly the same forms that they have now
3   that as the four biological systems occur only in living cells, the ancestor must have been a living cell
4   that all generations of living cells intermediate between the common ancestral cell and its living descendants had the same four biological systems in the same forms that they have now.

With ancient bones and bone, it was possible to confirm the predictions made about them by examining fossil bones and bone and cross-checking the results deductively. However when cells die, they normally decay rapidly and disappear and as a result rarely if ever fossilise. But even if they did fossilise and were available for examination, the predictions which have been made have been made about living cells not dead ones. Therefore the only way that the validity of the predictions made about past living cells can be assessed is by deductive reasoning.

Deductively, if the common ancestor of all living cells was a living cell, and all subsequent living cells have descended from it, that particular cell had to live successfully and it had to do so in a particular way. By definition, as it was the common ancestor of all subsequent living cells, it was not the descendant of any antecendent living cell. Therefore, when it came to life, it had to come to life suddenly, continue to live for an unknown period of time and at the end of its life, successfully initiate and successfully complete

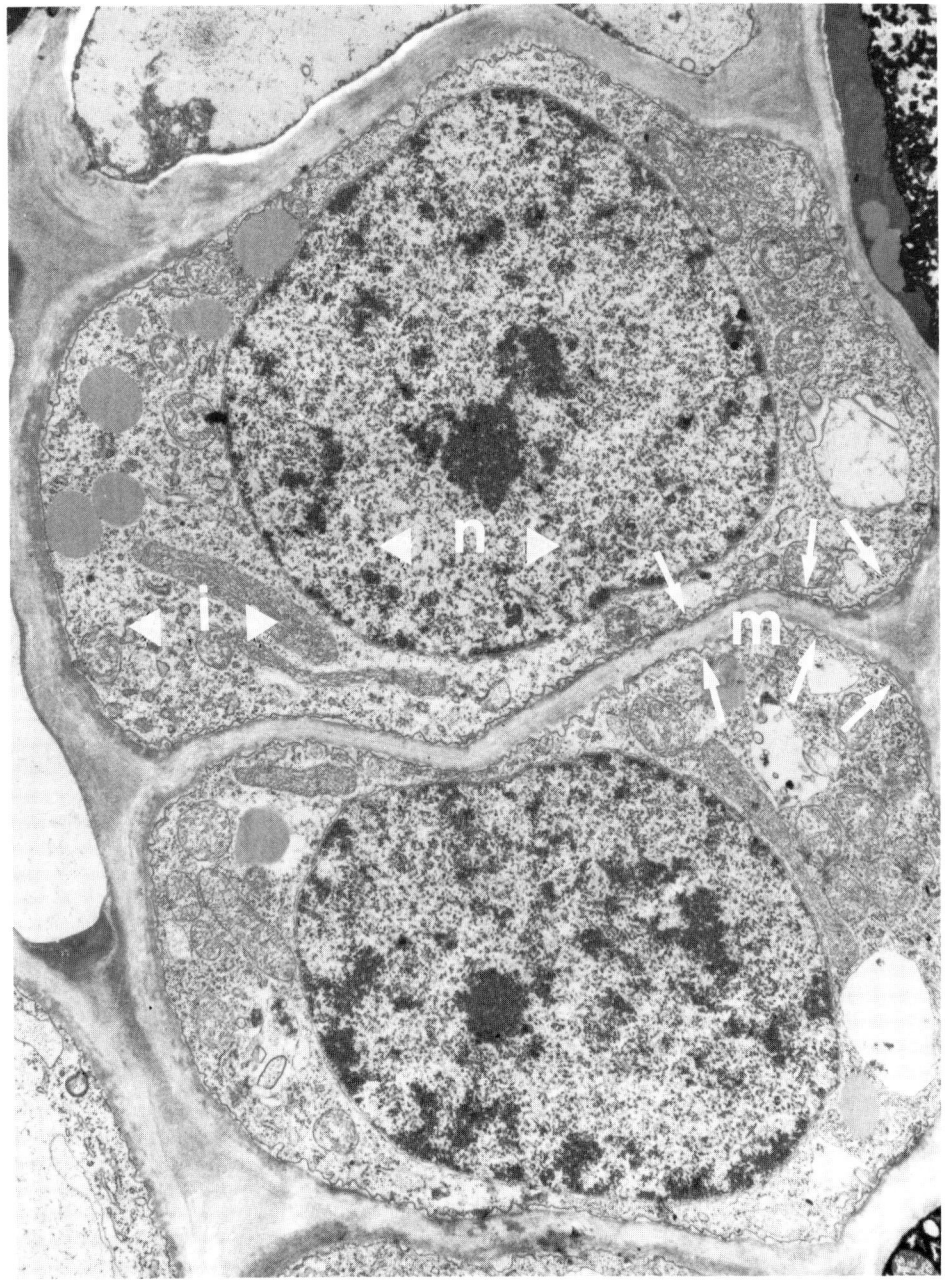

Cells.
An electron photomicrograph of two pinus radiata cells which have recently divided. The crenated cell membrane (m) forms the complete outer boundary of each cell. Within it there is the nucleus (n) containing much of the cell's DNA, and the intracellular content (i) which contains the energy-producing and energy-using systems of the cell. Immediately outside the cell membrane of each plant there is additional carbohydrate-based polysaccharide material which strengthens the cell membrane but does not replace it. The various bodies and spaces within the cell reflect its specialised living activities which are additional to its basic life. (magnification x 20,000)

its division into an immediate descendent generation of living cells. If this had not happened, it would not have established life on earth. But it did happen. The first cell must have lived successfully and it did establish life on earth and to do this it needed to have four basic characteristics or systems. It needed to have:

1 a complete outer cell wall otherwise its contents may have leaked away during its life or while it was dividing itself into its immediate living descendants.

2 an energy-producing mechanism which produced the energy it needs for its processes of living in a form which it could use.

3 an energy-using mechanism which used the energy to maintain the cell's vitality as it lived and reproduced itself.

4 a genetic control mechanism which masterminded all of these activities including initiating and completing its propagation into its immediate living descendants. As it propagated itself, the first cell's life and the essential biological systems of life were transfered into those descendants.

These systems were the essential elements of the first cell and despite the passage of time, they are still the essential elements of all living cells. All living cells still need to have a complete outer membrane to prevent their contents from escaping. They still need appropiate energy-producing and energy-using mechanisms for their processes of living and they still need a genetic control mechanism which masterminds those processes. Now however, the nature of those four systems is known. Loosely they have been called the family resemblances of cells, but precisely, they are the cell membrane with its specific structure; the anaerobic glycolytic system which produces ATP; the ribosome system controlled by RNA which uses the energy, and the DNA genetic control mechanism which controls the three systems. DNA also controls the other biological systems which may be present in particular cells, but regardless of them, the four basic biological systems, the family resemblances, are as essential to all successfully-living cells now as they were in the first living cell.

Although the exact nature of the DNA control mechanism has not yet been established, the nature of the other three systems has, and the only way that all presently-living cells can have exactly the same three highly sophisticated biological systems is for them to have inherited them from their common ancestor, the first living cell. The common ancestral cell must also have possessed the same systems in exactly the same forms as cells today otherwise all its presently-living descendants could not have the identical biological systems. For the same reason, all the generations of cells which have been intermediate between the first cell and its living descendants must also have had exactly the same biological systems functioning in exactly the same ways as they do now.

But all features of present life are controlled by genes and because three biological systems are the same in all living cells, it is reasonable to assume that their controlling genes, the fourth biological system, must also be the same in all living cells. Those genes must also be the same as those which were present in the first living cell and they must also be the same as those which have been present in all the generations of cells which have been intermediate between the ancestoral cell and today. All of the four systems are the same now as they were when they first appeared in the first cell of life, and by remaining so and persisting unchanged in all life, the genes associated with the systems have demonstrated that they are permanent genes.

There is some additional deductive evidence which supports this opinion and it becomes apparent once the nature of the life which is present in all living cells has been established.

Chapter 7

Life

Life is a biological process which, by definition, first appeared in the first living cell. Also by definition, it has also been present in all subsequent living cells. When the first cell suddenly came to life, in order for it to be alive and sustain its life, its four basic biological systems needed simultaneously to start to function in harmony and continue to do so throughout the whole of its life including the period of time when it was successfully propagating itself. The four systems were therefore intimately associated with the whole of the life of the cell. All subsequent successfully-living cells have also needed the same four systems functioning in continuous harmony throughout the whole of their lives and this means that within each of them, the same biological systems have always been intimately associated with the lives of each of them. Life is known to be masterminded by DNA but life is a dynamic process which requires and uses energy to maintain itself and as DNA is neither an energy-producing nor a major energy-using system, in itself, DNA is therefore not life. Similarly, life requires an intact cell wall to provide the necessary shelter for its living activities and DNA is not that wall. However, while life is being masterminded by DNA, DNA is also masterminding the actions of the three other biological systems which are essential to life. It therefore seems reasonable to postulate that the life which was present in the first cell was actually the combination of its four essential biological systems all functioning together in continuous harmony. It was not a fifth dimension of the cell. It was the continuous harmonious actions of its four systems. It is also reasonable to assume that the life which is still present in all living cells is still the same combination of those same four biological systems still acting together in the same continuous harmony. It is also the same life as that which was present in all earlier living cells and neither the life which was present in any living cells in the past nor any of its essential components have changed with the passage of time. This fits in perfectly with the earlier deduction that there are genes which were present in the first living cell which are still universally present in life today and still have exactly the forms that they had initially. They are the genes of cellular life and they are permanent genes.

It seems therefore that the neoDarwinian belief that the first spark of life appeared in a single living cell is correct. That belief has been derived from studies which have shown that there are family resemblances present in all living cells and that their universal distribution can only be explained if all life has stemmed from a common ancestral cell. However in the process of proving that this component of the theory is true, unfortunately for the theory, it has also been proved that the genes of those family resemblances are permanent genes. But permanent genes are an impossibility within the concept of the neoDarwinian theory and if it is accepted that they exist, at least part of the theory disintegrates. If that is so, it now becomes necessary to advance a new theory to explain the origin of life, but before that can be done it is necessary to investigate life a little further and also establish the form of the evolution which has actually occurred.

Cellular life, existence and death

As stated above, cellular life is the combination of an intact cell membrane, an anaerobic ATP production system, an ATP-using systems and their DNA control system all working together in continuous harmony and there are three consequences of that life being present in any cell.

1   *A cell may live successfully.*

To do so, at the end of its life it must successfully replicate itself into the next generation of living cells while simultaneously transferring its life's individual components and any additional characteristics it has into its living descendants.

2   *A cell may exist*

Successful living is controlled by each cell's own DNA. If that DNA does not have the ability to complete all the processes of successful living, although the cell may continue to survive for an indeterminate time, unless its DNA is restored to its full function and the cell successfully replicates itself, ultimately it dies. During the time in which it is not capable of living successfully, the cell may be said to exist, and existence is therefore the second possible consequence of a cell being alive. It is a normal biological state and many cells which exist are present in our bodies and presumably are present in all kingdoms of life (at least in all the multicelled kingdoms). For example, some of our nerve cells never reproduce themselves. If they retain the ability to reproduce and never exercise it, they remain fully alive until they die, but if the ability to reproduce somehow vanishes after the cell has formed, they exist. Most human skin cells exist. A layer of successfully-living skin cells in the depths of the skin continually propagate new cells on its superficial surface and each new cell displace older cells towards the surface of the skin where they ultimately die and disappear. They never reproduce and it seems that even at the time of their formation, this ability may be deficient. They just exist from the time that they are formed until they die.

Gametes, mature sperms and ova, exist. When they are formed they lack their full complement of the four biological systems of life and this prevents them from living successfully. Instead, they continues to exist until they die or until they unite with other suitably-prepared gametes. Such fusions restore the four systems of each of the resulting composite cells to their required levels and the composite cells then demonstrate that they have been restored to full life by successfully dividing themselves into their next generation of living cells.

3 A cell may die

Death is the third possible consequence of cellular life. Death occurs when the cell's four biological systems cease to act together at an acceptable level of continuous harmony and decay sets in. Decay is an essential part of death and the onset of decay indicates that death has occurred. However death does not always occur when the levels of the biological activity of the four systems fall below their normal levels. For example, a chilled cell may be alive even though the levels of its biological activities may be almost nil. Such a cell may remain alive in a state of suspended animation until either its biological activities return to normal when the temperature of the cell returns towards normal, or decay sets in and death occurs.

Corporate life and death

The life which is present in living multicelled organisms is more than the summation of all the cellular life which is present in their individual living cells. Multicelled organisms have an additional form of life, corporate life, which controls the living processes of the organism as a whole and directs its collective living activities. Corporate life does not seem to control the cellular life of individual cells. That seems to remain as an independent activity of each cell under its own DNA control but corporate life does direct the activities of cells other than those of their basic lives. Corporate life also has the capacity to direct the changing of cellular life to cellular existence as occurs in gametes. As with cellular life, corporate life is a genetically-controlled biological process with its genes residing in cells. The genes are not universally distributed throughout multicelled organisms and the percentage of the cells of the organism which contain corporate life genes varies from species to species. In the animates, the corporate genes tend to be more restricted in their distribution than in plants and fungi, and in animates in particular, it is evident that the more complex the species, the more limited is the distribution of their corporate life genes in their bodies. It is not known whether corporate life genes are identical in the three multicelled kingdoms though it seems unlikely that they are. If there was much

Life around us.

Life slowly ebbing away.

The rapid onset of death.

common identity between all corporate genes, life should be much more similar in the three multi-celled kingdoms than it is.

Corporate death is the inevitable consequence of corporate life. It occurs when sufficient numbers of the cells housing the corporate life genes have been compromised and corporate life ceases. Decay then occurs and the state of corporate death is established. Individual cells may continue to live for some time but eventually they also die.

Corporate existence either does not or cannot occur in complete organisms though a form of corporate existence can occur in some organism's individual components and organ transplanting takes advantage of this. Organs of the body have a limited form of corporate existence which is more than the summation of the cellular life of their individual cells. When an organ is transplanted successfully, the corporate life of the new host is able to replace the absent corporate life of the donor and take control of its proper healthy performance. However the consequences of the organ existing outside its mantle of corporate life are the same as the consequences of the existence of the organ's cells. Unless the organ is restored to the protection of corporate life both it and its individual cells die.

Viruses

Viruses are not cells and because of this, they are not alive. They exist. Although they all contain parts of one or more of the four basic biological

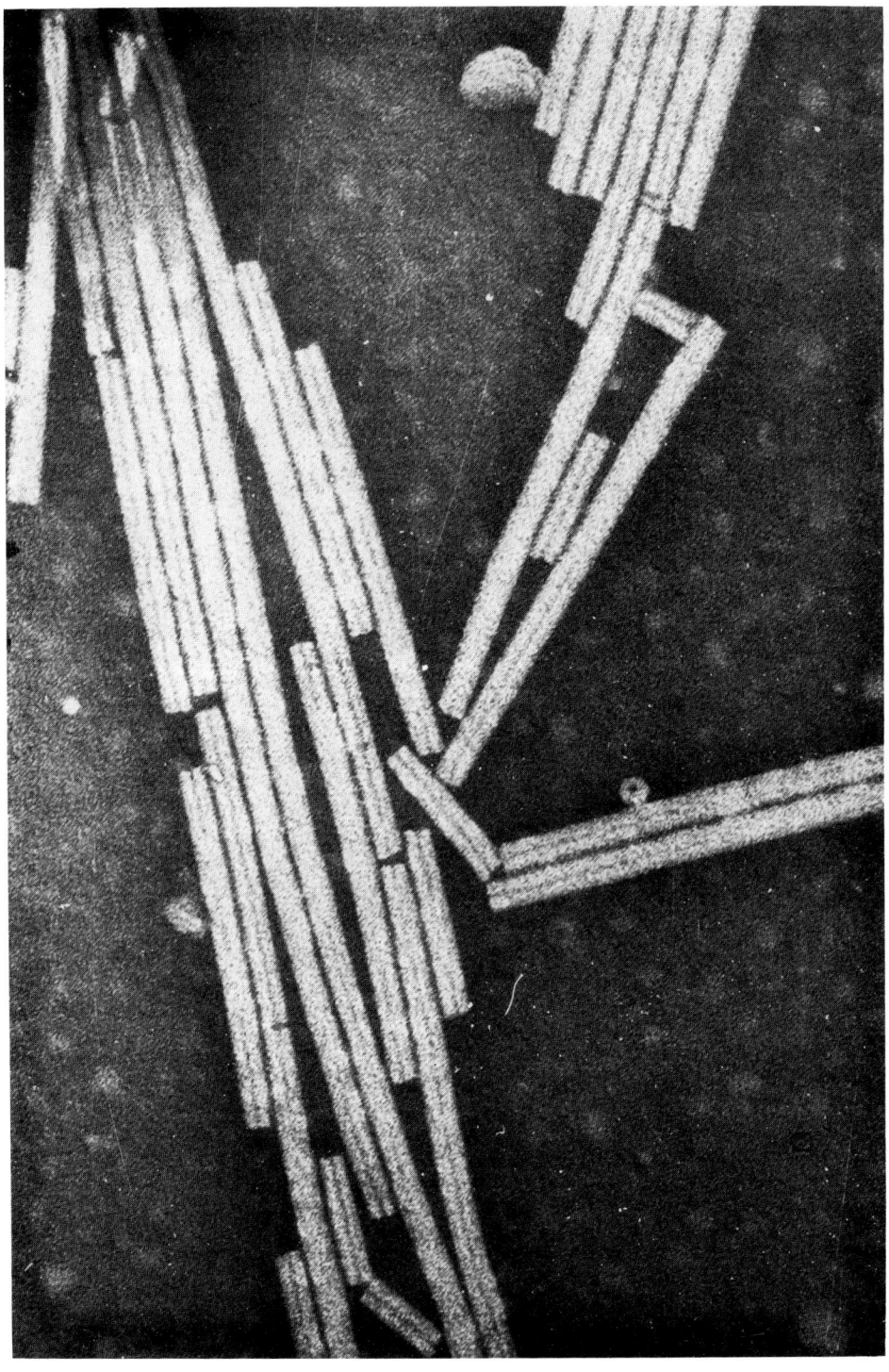

Tobacco mosaic viruses.
These lack the structured form of cells and are obligatory intracellular inhabitants.
(magnification x 431,000)

systems of life, none have all of the four systems in their complete forms functioning in continuous harmony. If they did, they would be living cells and not viruses. For this reason, viruses can never be alive.

To maintain their existence, viruses need aspects of the four basic biological systems which they lack and in order to obtain them they must reside within living cells. Particularly, they need to use the host's genetic system to reproduce themselves for although they all have some general form of genetic system, and in some instances seem to be composed of nothing else but DNA, none has the ability to initiate and complete its reproduction without the help of its host cell's genes.

Because all viruses lack some of the essential biological elements of life, their ancestors must also have lacked some of those elements. If that is so, viruses have never been alive, but as they all share some of the family resemblances of living cells, it seems certain that they are descendants and not antecendents of the first living cell. If they were antecendents, their universal dependence on living cells for their existence would be difficult to explain. However, as descendants of the first cell, it is logical to surmise that some stage, one or more living cells lost parts of their biological systems and simultaneously successfully managed to parasitise other cells for their needs.

A comparison of cells, gametes and viruses

	Cells	Gametes	Viruses
Complete, biologically-active cell membrane	yes	?yes	no
ATP production by anaerobic glycolysis	yes	?yes	?
Ribosomes controlled by RNA	yes	?yes	?
DNA with demonstrated ability to initiate and complete reproduction	yes/no	no	no
Dead	no	no	no
Fully alive	yes(some)	no	no
Existing	yes(some)	yes	yes

(When a pair of appropriate gametes fuse they form a cell which is alive.)

As with the stable genes of cells, viral genes are also presumed to have their own levels of stability and if these are comparable to cells, it would seem likely that there are core genes of viruses which are extremely stable and there are also a minority of genes which have the ability to change with varying degrees of flexibility. Changes to them produce the species differences which distinguish similar species of viruses from one another.

Chapter 8

The evolution which has actually occurred

Evolution has occurred but not in the illogical and haphazard way it should have had it happened in the neoDarwinian way. Instead, living and fossil evidence shows that evolution has followed two complementary and predictable pathways. Well-defined sequences of key related species of increasing complexity have appeared progressively in all the kingdoms of life and as each has appeared, its appearance has increased the complexity of life on earth. At each new level of complexity in each of the ascending sequences, streams of related species of comparable degrees of complexity have stemmed from these key species and these descendant streams of species have produced the diversity of life. A study of the evolution of the domestic dog illustrates this.

The Dog

Kingdom	Animalia
Phylum	Chordata
Subphylum	Vertebrata
Superclass	Tetrapoda
Class	Mammalia
Order	Carnivora
Family	Canidae
Genus	Canis
Species	Canis familiaris

The chart shows that the dog is part of the Animal kingdom and as the ancestral species of that kingdom was the first multicelled animate species which appeared, the dog's oldest multicelled ancestor is that particular key species. As has already been shown, because it is possible to study past life forms by studying their living descendants, that key species can now be studied in some detail by examining living animates, locating features which are universally present and identical in them and projecting these back into that key ancestral animate species. Similarly for the other steps which are recorded in the chart. The phylum Chordata started in one specific key species and it and some of the evolutionary events which surrounded its appearance can also be studied. So too can the ancestral species of the subphylum Vertebrata, the ancestral species of the superclass Tetrapoda, and all the other ancestral species of increasing complexity which are recorded in the chart. Each step was the result of the appearance of a specific key species and as each one appeared, life on earth became more complex and ascended. Once the nature of those species and the nature of some of the genetic events which must have been associated with their appearance have been established, it is then possible to determine the genetic changes which resulted in the appearance of the streams of species of comparable complexity which stemmed from them and diversified life.

There are a number of features which are universally present and identical in all presently living members of the Animal Kingdom including most aspects of their protein-based intercellular glue(s) and most aspects of their collagen fibres. This indicates that when the ancestral animate species appeared approximately 1000Myrs ago, it possessed the features in the same forms that they have today. Subsequently, as variations occurred to aspects of their animate features, a stream of new animate species of comparable complexity to the original animate species is presumed to have progressively appeared. Where the resulting variations were of sufficient magnitude, those which we would classify as new species appeared, and where the variations were of a lesser magnitude, those which we would classify as varieties of existing species appeared. Though the appearance of these animate species and their varieties increased the diversity of the existing animate life on earth, because they all contained the same core features as their ancestral species, they did not increase the complexity of life.

At some later time, a new species appeared which had a pattern of features which was more complex than those of the existing animates and these new features established that it was a chordate species. However, the new features did not replace the animate features. They were additional to them and the possession of both sets of features established that at one and the same time, it was both a chordate and an animate. It was not either one or the other. It was both and with its appearance, the phylum Chordata had started and evolution in the animate kingdom had taken a step upward. A stream of chordate species of comparable complexity to it is then presumed to have appeared as variations occurred to the combination of animate and chordate features and the new stream of chordate species appeared parallel to the existing stream of animates. These two streams increased the diversity of animate life without changing its now dual level of complexity.

In time, the next step in the evolution of the dog occurred when the bony vertebrate pattern of features appeared in a chordate species superimposed on its existing animate and chordate features and with that, the ancestral species of the subphylum Vertebrata had come into being thereby increasing the level of complexity of life. As variations occurred to its combination of animate, chordate and bony vertebrate features, a new stream of bony vertebrate species appeared parallel to the existing streams of chordates and animates which had the same degree of complexity as their key ancestral bony vertebrate species. Their appearance added to the diversity of life.

The next important evolutionary step was the appearance of a species which had the pentadactyl tetrapodal vertebrate (PTV) pattern of feature superimposed on the animate, chordate and bony vertebrate features of its parent species. This new and more complex species was the ancestral species of the superclass Tetrapoda, and as variations began to ocur in its combination of features, another stream of new species with its degree of complexity began formed parallel to the existing streams of species of lesser complexity.

The next key ancestral species of the dog which appeared was the ancestral species of the reptiles, an important ancestor which is not usually included in Animalia classification charts. It had its own particular level of complexity and it spawned its own particular stream of new reptilian species. In time, in one of these, the ancestral species of the class mammalia appeared with its own distinctive pattern of features superimposed on the existing patterns of animate, chordate, bony vertebrate, PTV reptilian features and it also spawned its own stream of new species of comparable complexity parallel to the existing streams. Further key ancestor species of steadily increasing complexity for the order Carnivora and the family Canidae appeared in sequence until finally, the ancestral species of the genus Canis came into being with its own distinctive pattern of features superimposed on the core patterns of all the others in the preceding sequence of key ancestral species. It also spawned its own stream of new species of comparable complexity and in one of them, a particular cluster of

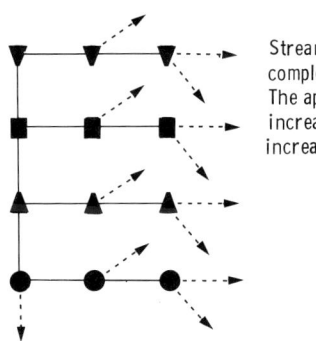

Streams of new species of comparable complexity to their key ancestor.
The appearance of each new species increases the diversity of life without increasing its complexity.

A sequence of key ancestors of increasing complexity.
The appearance of each new one marks another step in the ascent and the complexity of life.

The tree of life (in a very stylised form)

modifications occurred to its essential features. With their development another new species had appeared which had particular interest for us and at last, *Canis familiaris* had appeared on earth.

Classification charts similar to that of the dog show that similar sequences of evolution have occurred for all species in all the kingdoms of life since life began. Descendent sequences of key ancestral species of progressively increasing complexity have appeared from time to time each containing an accumulation of the patterns of features of earlier key ancestors plus a new more complex pattern. Streams of species of equal complexities have then stemmed from them as modifications have occurred to their essential features. If these modifications have been of sufficient magnitude, separate species have formed and if lesser changes have occurred, varieties of the same species have been formed. If variations which have formed, as they are always doing, and they have been only of a very minor nature, the results have formed the kinds of changes to individuals within species which Mendel observed occurring in his flowers.

The genetic changes which have produced evolution

If the combination of the appearance of key species of increasing complexity and the streams of equal complexity which have stemmed from them has been evolution, and if the genetic changes which have produced these two types of species can be defined, the genetic processes of evolution will have been established. To determine this, the most important sequence of key species which can be examined are those which preceded and initiated the appearance of the Devonian amphibian PTV ancestor. Once the nature of the evolution of these key species has been discerned, the stream of PTV species of comparable complexity which streamed from it can then be studied.

Studies of living animates have shown that similar features are present in all of them and within each feature there are elements which are different and elements which are the same. They are the same because they have been inherited in their present forms from their common ancestor and by projecting these elements back into that ancestor some of its basic structures can be established. This must also establish some of the ancestor's basic genetic makeup.

The genes of the elements which are identical in all living animates must have been present in the ancestral animate species in their present-day forms otherwise all living animates could not share the identical features which they do share. This also means that the genes have remained the same since the ancestral animate species first appeared. Therefore if the ways that evolution has occurred are going to be explained, the question of how those genes appeared in that animate ancestor in the first instance must be answered. Did the genes come from an earlier single-celled ancestor or were they new genes which appeared for the first time in life when the first multicelled animate species came into being 1000Myrs ago? The answer can be obtained deductively.

While some of the original animate genes are universally present in all living animates, they appear to be universally absent in all other species in all the other kingdoms of life. If this is false and there are some of the original animate genes in some species in the other kingdoms of life, they are present in them either because they were present in a life form which an ancestor of both the ancestral species of the animates and of the other kingdoms of life, or they were passed on to the other kingdoms by animates after they had appeared in the ancestral animate species. Therefore if there are essential animate genes in any of the other kingdoms of life it is not possible to determine how they appeared in the first instance, but if those core genes are completely absent from the other kingdoms of life and are universally present in all animates, it is possible to make an accurate guess as to how and when they appeared. For them to be universally present only in the animate kingdom, they had to come into being only as the kingdom's ancestral species came into being otherwise they could not be universally present today in the animate kingdom in the way that they are and universally absent in the other kingdoms of life.

The chordate genes must have appeared in this way. Because chordate genes are present only in chordate animates, when the new chordate features and the genes which formulated them appeared in the animate species which spawned the new species, the chordate genes must have come into being as the chordate species itself came into being.

The exact nature of some of these suddenly-appearing genes can be determined. Chordates are so named because they all have a notochord and studies of living notochords show that they are all macroscopically and microscopically the same except for the presence of some minor species differences. They can only be universally the same in all chordate life today if their controlling genes are also universally the same. For this to be so, the genes must also have been present in exactly the same forms that they have now in the ancestral chordate species. There is no evidence that a notochord (or any of the other essential chordate features) are present in other parts of the animate kingdom or in any of the other kingdoms of life. Therefore the chordate genes were new genes when they first appeared in the ancestral chordate species. When they appeared, they came into contact with existing animate genes and seem to have modified their activities without actually changing them. The resulting combination of animate and chordate genes have subsequently remained extremely stable except for the effects of some changeable genes amongst them. These changeable genes have shown that they can change and when they have done so they have produced the species differences which are present in the core chordate features today.

When the bony vertebrate ancestral species appeared, the genetic activities surrounding its appearance followed the same pattern. While essential animate and chordate genes persisted in the new bony vertebrate species, a new set of genes suddenly appeared in association with them. The new genes interacted with the existing genes modifying their activities without actually changing them and this combination of animate, chordate and bony vertebrate genes fashioned the new more complex ancestral species of the bony vertebrates. As with the genetic combinations of its key antecendents, the genetic combination of the new bony vertebrate species subsequently remained very stable.

It is not clear if some existing genes vanished as the new genes appeared though it seems likely that they did. It seems to have happened when the ancestral PTV species of the Devonian came into being. Its fossil and living descendants show that when it appeared, some of the core genes of its immediate antecendent species (which must have been a fish as it was, by definition, the first pentadactyl tetrapodal vertebate species) passed into it; some of the antecendent's genes vanished and equally as suddenly, some new ones appeared.

Using the triangle of study to study the ancestral species of the pentadactyl tetrapodal vertebrates, studies of its living descendants show:

1   that it possessed the genes of the features which are common to all living animates today,

2   that it possessed the genes which are common to all living chordates today,

3   that it possessed the genes which are common to all living bony vertebrates today,

4   that it possessed the genes of bone which are common to all living bony species today,

5   that it possessed the genes of the rigid PTV bones which are common to all living PTV's today,

6   but it did not possess the genes of rigid fish bones which are common to all living bony fish today.

Yet the sequences of species which lay between the ancestral bony vertebrate species and ancestral pentadactyl tetrapodal vertebrate species were all bony fish and the nature of some of the genes of these species can be established by studying their living descendants. The author's limited radiographic study of bony fish supports the generally held belief that all fish bone is the same and is the same as all PTV bone. It also showed that as far as could be observed, all fish endoskeletons are the same (subject to the species differences which are present). If fish skeletons are the same in all living fish, their genes must be the same as must have been the fish skeletons and the fish skeleton genes of their ancient ancestors. Therefore, when the PTV ancestral species came into being on the basis of an existing bony fish species, the new PTV species possessed that species core animate, chordate and bony vertebrate genes and its core genes of bone, but it did not have its core genes of fish bones. In their place there were new genes which formed the new species bone tissue into PTV bones.

It may be argued that the fish skeleton genes did not actually vanish as the new PTV species formed and they just changed into PTV bones' genes. This possibility cannot be denied but the more likely explanation for the abrupt increase in complexity which occurred in the PTV ancestral species, is that some of the old genes vanished and some new ones appeared in their place. The fossilised bones and bone of the PTV ancestor support this view, but regardless of whether some genes vanished and some new ones appeared, or whether one set of genes was transformed into another, the changes occurred very rapidly and they also occurred irrevocably. From then on, the new genes demonstrated that they had and still have the same degree of stability as the existing animate, chordate and bony vertebrate genes the ancestral PTV species received from its fish ancestors.

Key ancestors in the other sequences of increasingly more complex key ancestral species in all the kingdoms of life are presumed to have followed the same pattern of genetic movements as they came into being. They retained most of the existing core genes of their immediate antecendent species; some of the antecendants' genes disappeared, and in their place some new ones appeared. The new genes then modified the actions of the persisting genes without actually changing them, and these new genetic mixtures with their associated new more complex pattern of features, were then inherited by subsequent descendants.

If these changes were the ways that life on earth became increasingly more complex, the genetic changes which resulted in the increase in the diversity of life occurred in a different way. Each new stream of species which flowed outwards from the key ancestral species received its core genes, but amongst the genes were some which had the capacity to change. When they did change, they modified the actions of the remaining genes, stable and changeable, and the result was the formation of a new species with the same degree of complexity as its key ancestral species but with a slightly different set of features. The ways that the PTV skeleton has changed in the last 400Myrs demonstate this.

Since it appeared in the Devonian, the PTV skeleton has effectively remained the same, subject only to the limited number of species differences which have appeared. The number of bony PTV skeletal genes which have masterminded the differences appear to have been small and their activities seem to have been specific. It seems that only relatively minor changes to their activities have been sufficient to modify the actions of their associated stable genes and to formulate changes which are of sufficient magnitude for us to label their possessors as being part of different species though the species themselves have the same degree of bony complexity as their predecessors.

All changeable genes do not have the same degree of susceptibility to change and this susceptibility seems to range from being almost stable to being very labile. Where changeable genes at the more stable end of the spectrum have changed and have modified the actions of other stable and

78 EVOLUTION AND A CREATOR?

labile genes, species or perhaps varieties of species have appeared, but where the more labile genes have changed, only minor or relatively minor changes have occurred to the features controlled by stable genes (Mendel's flowers again).

The great diversity of life is a reflection of the activities of changeable genes acting on stable gene combinations and on other changeable genes. The great complexity of life has resulted from the sudden appearance of those genes. Both the stable genetic combinations and the changeable genes seem to have come into being simultaneously in the key ancestral species but what has not yet been established in this discussion is what or how many individuals were actually involved in these genetic revolutions and at what stage in their lives did the changes actually occur.

The only time in the life of any individual organism when changes to genes will be transmitted with certainty into its next generation, is as the first cell of the new organism is receiving its genes from its immediate forebear(s). Genetic changes occurring at this time affect all of its subsequent cellular development and are therefore passed on to its next generation. Genetic changes which occur after this do not necessarily change the form of the organism, nor are they necessarily transmitted into its next generation.

It is generally accepted that new species stem from only a very limited number of changed individuals. However, because of the exactness of the genetic changes which appear to have occurred in the key ancestral species, it seems that only one or two individuals, at the most, can have been involved in generating the appearance of a new species. The identical nature of genes in all the descendants of those individuals hundreds of millions of years later, does not permit many to be involved in the formation of the new species. If the new genetic formats had developed gradually in the creeping way which is inherent in the neoDarwinian theory of evolution, over sequences of generations, slightly different mixtures of the stable genes would have formed as organisms with slightly different genetic combinations interbred. Subtle or perhaps not so subtle differences would be therefore be present today in the features controlled by those different but similar genetic combinations. But they are not present. Instead, identical features and therefore identical genes are present hundreds of millions of years later in species which are now at best, only very vaguely related to each other, and their presence infers that only one individual was the foundation member of the species if the species reproduced asexually, and two if the species reproduced sexually.

It would seem that in asexually-producing species, as one individual was in the process of being generated from its immediate forebear and as the forebear's DNA was being passed into it, some of that DNA vanished and some new DNA appeared in its place. The resulting organism then possessed a new genetic combination which made it a more complex individual and it became the foundation member of a new species. Its new genetic combinations were then inherited unchanged by its descendants and because of the stability of the combination, the pattern of features which they controlled have persisted in the same form ever since subject only to the actions of changeable genes.

In the sexually-reproducing species, a very similar sequence of events must have occurred as new key individuals appeared with the added complication that the genetic changes had to occur simultaneously in at least one of each sex. The changes must also have been of such a magnitude that the new key individuals and their descendants could not mate successfully with members of the antecendent species nor could they mate with members of different species. Had this been possible, many different stable genetic combinations would have formed forming many similar key individuals and many similar key species. The species classification charts however indicate that this did not happen and only one key species formed in each descendent sequence.

Conclusion

The study of evolution was undertaken to determine what kind of evolution followed the appearance of the first cell of life and it appears from available living and fossil evidence that it has followed two complementary pathways. Whatever the kingdom which has been involved, new genetic combinations have suddenly appeared in one or in two organisms and the resulting changed organisms have founded a new more complex key species. The genetic combinations in these species have then remained stable in their descendants and any subsequent changes to them have been effected by the actions of a limited number of changeable genes which also appeared in key individuals. Depending on the degree of lability of these changeable genes, their actions have produced new species, new varieties, or the minor variations which are always appearing in the individuals of species.

Chapter 9

Theories about the origin and the evolution of life on earth

Three major theories have been advanced in the past to explain the origin and the evolution of life on earth. The first was the Bible's account of the Creation. When it was in vogue it was accepted that rather than God having created a fully-populated world, He created examples of every species and directed them to populate the earth by their own actions. It was also understood that because God was perfect, He had created perfect creatures and as such they were therefore unchangeable. However when Darwin found that this was not true and showed that some species had changed with the passage of time, scientific and later theological belief in the truth of Bible's theory vanished and Darwin's theory was accepted in its place. He theorised that life on earth had evolved and had done so in a particular way, but while living and fossil evidence confirms that life has evolved, it does not support the belief that it has evolved in Darwin's way. Instead, the evidence supports the belief that evolution has occurred in a different way and if that is true, it seriously compromises the neoDarwinian version of his theory. Observations also indicate that some permanent genes exist in life and if this is true, as such genes cannot exist within the neoDarwinian theory, it collapses, as with the Biblical theory before it.

The third major theory which has been advanced from time to time to explain the evolution of life is the "offshore" theory. In its several forms it holds that the origin of our life and/or the origin of some of the evolutionary changes which have occurred and/or the origin of some of the diseases which have swept the earth from time to time have been the result of seedings of life or fragments of life from somewhere out in space. Aspects of this theory have recently been supported amongst others by a sophisticated mathematical analysis made by Edward Anders but there is evidence against this theory.

Because of its fiery origin, the world was sterile at its inception, therefore if any seedings of life rather than fragments of it have come to us at different times from space, each has required its own separate transport system. Each successful seeding must have successfully traversed the intense cold of space; survived the descent through our atmosphere; landed in an hospitable place; freed itself from its transporter and united successfully with existing life forms.

As all our life is cells and as they all share the same family resemblances, if our life started as the result of a successful seeding from outer space, it can have happened only once. Also, as there is fossil evidence that life was present on earth 2700Myrs ago, it must have happened before that. Therefore if there have been a number of seedings, either that very ancient one is the only one which survived, or they have always stemmed from the same ancestor from the same space source. Had they stemmed from more than one source in space or had they stemmed from more than one ancestor at that source, our present cells could not all share the same family resemblances.

If our life has come from the same ancestor in space and our evolution has been the result of progressive cellular seedings from it, some guesses can be made about the life form of that ancestor. If it successfully seeded us with cells containing genes which led to the relatively abrupt major developments in evolution such as photosynthesis, sexual cell division and the original Devonian pentadactyl tetrapodal vertebrate (PTV), then its own evolutionary development must at least be similar to that on earth and perhaps millions, or even hundreds of millions, of years in advance of us. To have been responsible for life as we know it on earth, the biology and biochemistry of the ancestor source in space must be somewhat similar to ours and the one trace of it, or them, which may be discernable at a vast astronomical distance is the presence of atmospheric oxygen. No evidence of atmospheric oxygen has yet been found in space despite almost constant searching by astronomers for many years.

It may be argued that the seedings have not been cellular seedings but have been fragments of life or viral seedings and that such seedings have been responsible for the sudden appearance of stable genes in new key ancestors or have been the cause of infectious scourges which have swept the world from time to time. The reasoning against this is similar to the reasoning against cellular seedings. As DNA is the architect of all the life which is present on earth and is also the architect of infectious diseases, any alien viral seedings will have needed to have contained DNA which has been compatible with earthly DNA otherwise the alien viruses may not have been able to enter our cells and successfully exist there. If however, our viral population is a mixture of earthly and alien viruses, as all viruses appear to share common family resemblances, both the alien viruses and the earthly viruses must have come from the same ancestor on the same source otherwise all viruses could not share our family resemblances in the way that they do. Earthly viruses however are not capable of independent existence and if they are not offshoots of our earlier life and have come from space, unless they have had different characteristics in space and have been able to exist independently, they will have needed living cells to host them during their travels. There is no evidence of any of those cells on earth. The possibility that we have been seeded with viruses is as unlikely as is the possibiltiy that we have been seeded with cells.

The various versions of the "offshore" theory do not make sense when they are examined against the known structure of present and past life and unless some very compelling evidence for the theory is discovered in the future, it can be discounted as being relevant to our life or its evolution.

Any new theory which is now advanced to explain the abrupt and massive genetic changes which have been associated with the appearance of key individuals and their associated key species, and to explain the more subtle genetic changes which have formulated their descendant new species, new varieties and the minor variations which are always occurring, must first answer the criticism that because other life forms have always been near by when they have formed, all the genetic changes which have occurred have been the result of amalgamations of the genes of some of these life forms. This criticism cannot be effectively answered until the way that the first cell of life appeared is satisfactorily explained. When that has been done, that explanation may then be extrapolated into the events which surrounded the appearance of subsequent forms of life.

It has been shown that the first cell of life was a combination of four very complex biological systems which came into continuous harmonious interaction with each other from the first moment that the cell began to live, and continued to act in that way throughout the whole of its life. Three of the systems were controlled by the fourth system, the genes, and these genes have been shown to have remained the same in all living cells ever since. If the "offshore" theory which has been propounded has been successfully discounted, there are only two possible ways that those four systems can have appeared in the way that they did. Either they appeared by chance or they did not appear by chance. The subsequent permanence of the genes which were in that first cell, and its incalculable number of descendants, and the considerable complexity of their subservient biological systems must go a long way towards eliminating chance as being a reasonable explanation for the cell's appearance. Therefore if chance was not its architect, one must now consider the possibility that some other agency must have formed it. Such an agency would have had to have been a "genetic engineer" on a large scale. On a small scale, we are already capable of some genetic engineering, but this "engineer", to be effective, had to have the ability to design the whole cell and successfully complete its formation.

If such an agency did form the first cell with all its immense interlocking complexity of form, the lesser genetic engineering changes which have been necessary to formulate the various subsequent key ancestral organisms which have appeared from time to time becomes much more understandable. As with the appearance of the first cell, those changes do not seem to have happened by chance. They have been too complex and have occurred too abruptly for that. There is also evidence which argues against their appearance having been the result of interactions between existing genes in adjacent life forms. Every stable gene which is present in one segment of a kingdom of life and is absent from every other segment of the same kingdom denies that this process occurred. Genetic uniqueness and the great genetic stability which is present in all aspects of all life cannot be explained these ways. It is much more logical to assume that if the first cell was deliberately genetically-engineered, the same agency also deliberately genetically-engineered most and perhaps almost all of the subsequent genetic changes which have occurred during the evolution of life. What is not completely clear is whether the "genetic engineer" has been the only agency involved in evolution. Chance, natural radiation and any other agency which can change genes may also have been involved in changing some of them at some time, particularly those changes which have occurred to the more labile genes. All of them may have been involved to a greater or lesser extent in some of the lesser aspects of evolution, but their involvement appear to have been minimal compared to the involvement of the genetic-engineering agency in the appearance of the first cell and its involvement in the subsequent genetic steps of evolution. Little of the evolution which has occurred, including the origin of life itself, can reasonably be explained as having been the result of randomly occurring genetic changes with natural selection of the results, whereas most of the genetic activities which have been necessary to bring the first cell of life into being and effect the subsequent steps of evolution can be explained as being the actions of a precise "genetic engineer".

If that is so, has evolution actually been the result of the actions of a Creator?

Bibliography

Many texts and articles have been studied as part of the preparation of this book and some of the most useful have been:

The Holy Bible

The Origin of Species
Charles Darwin. 1st edition 1859.
Edited by Burrow J. W., Penguin Classics 1959.

Origin of Species
Charles Darwin. 6th edition 1882.
Edited by Leakey R. E., Oxford University Press 1977.

The Origin of Species by Charles Darwin – A variorum text.
Edited by Morse Peckham, University of Pennsylvania Press. 1959.

Darwin – An illustrated life of Charles Darwin, 1809–1882. 2nd edition.
Fletcher F. D., Shire Publications Ltd. Aylesbury 1988.

The Molecular Biology of The Cell
Alberts A., Bray D., Lewis J., Raff M., Roberts K., & Watson J. D., Garland Publishing, Inc. New York & London. 1983.

Vertebrate Paleontology and Evolution
Carroll R. L., W. H. Freeman & Company. New York. 1988.

A Textbook of Histology – Bloom and Fawcett 11th edition.
Edited by Fawcett D. W.,
W. B. Saunders Co. Philadelphia & London 1986.

Hen's Teeth and Horses Toes
Gould S. J., Penguin Books, 1983.

The Neck of the Giraffe
Hitchings F., Pan Books Ltd. London 1982.

Principles of Geology
Lyell C., John Murray, London 1830–3.

Five Kingdoms – An Illustrated Guide to the Phylla of Life on Earth. 2nd edition. Margulis L. & Schwartz K. V., W. H. Freeman & Company, New York 1988.

A View of the Evidences of Christianity 23rd edition.
Paley William, John W. Parker & Son, London. 1859.

Fish Pathology. Roberts R. J., Editor. Bailliere Tindall London 1978.

The Vertebrate Body 5th edition.
Romer A. S., & Parsons T. S.,
W. B. Saunders Company, Philadelphia, London 1977.

The Great Evolution Mystery
Taylor G. R., Abacus, London, 1984.

Prebiotic organic matter from comets and asteroids
Anders E., *Nature* 324 : pp255-257, 1989.

The oldest fossil
Baaghorn E. S., *Scientific American* 224(5): pp30-42,1971.

Evolution of Calcium Regulation in Lower Vertebrates
Clark N. B., *American Zoologist* 23: pp719-727, 1983.

The relative number of living and fossil species of animals
Miller S. W. & Campbell A., *Systematic Zoology* 3: pp168-170, 1954.

The most ancient rocks revisited
Moorbath S., *Nature*, 321: p725, 1986.

Quelques Remarques sur L'Histore Evolution Des Tissues Sequelettique Chez Les Vertebres et Plus Particulierement Chez Les Tetrapods.
de Riqles A., *Annee Biologique* 18(1-2): pp1-35. 1979.

The oldest eukaryotic cell
Vidal G., *Scientific American* 250(2): pp32-41,1984.

The origin of the tetrapods
Westoll T. S., *Biological Reviews* 18: pp78-98, 1943.

Wickramasinghe, Professor Chandra & Hoyle, Sir Fred.
Sunday Times p3, 10.12.1989.

Glossary

aerobic glycolysis – the form of glycolysis which needs oxygen for its action. It is more efficient and produces more energy per unit of sugar than anaerobic glycolysis does.

anaerobic glycolysis – the form of glycolysis which does not need oxygen for its action. It is a much less efficient energy-producing system than aerobic glycolysis.

articulate – divided into or united by joints

ATP – adenosine triphosphate. The energy-rich biological substance which all organisms use for their energy needs

avian – pertaining to birds

bone – the combination of the rigid tissue which forms bones and a complex dynamic biochemical system

bones – rigid support structures of a body which have been formed from bone

biology – the science of life

biotic – relating to life or living things

cell wall – the outer boundary of all living cells

chordates – animate species which have notochords

collagen – a protein-based fibrous substance of animates which may range in form from being very small strands beside individual cells to large tendons beside joints

cyanobacteria – believed to be the first bacteria to have the power to photosynthesise and produce oxygen. They are still present on earth, but today they are only a minor group of bacteria

Devonian – the period of time between 425 Myrs and 375Myrs ago. During this time, the land became populated by plants and animates including amphibians. Prior to the Devonian period, of the bony vertebrates, only fossils of fish have been found, but from the Devonian period onwards, both fish and PTV fossils have been found and it is this which establishes when PTV's first appeared.

digits – fingers and toes in humans and their equivalents in other mammals, birds, reptiles and amphibians.

disarticulate – separate bones from one another through their connecting joints (articulations)

DNA – desoxyribonucleic acid. It is present in cells mainly in the form of strands and it is the biological substance which forms genes

endoskeleton(bony) – the skeleton which is within the body of a bony organism

exoskeleton (bony) – the bony skeleton which is within the skin of an organism or is part of its surface. The bony plates of dinosaurs, the bony plates of armadillos and the antlers of deer are examples of exoskeletons. The bone which constitutes them is the same as that of endoskeletal bones and in most instances, it remains biologically active throughout the life of the organism. Athough bony exoskeletons may have constant forms in a particular species, there is no consistency in their forms in different species

exoskeleton (non-bony) – the support structures formed by an individual organism which are outside its body. The hard shell of a crab and the shells of shellfish are examples of exoskeletons. Although non-bony exoskeletons are the results of cellular activities of the organism, unlike bony exoskeletons, they are inert

flagellum – a whip-like structure which protrudes from some cells. The cell can move the flagellum and in some cells, the movement of the flagellum moves the cell itself

gamete – either a mature sperm or a mature ovum which can unite with the other to form a new organism.

gene – a unit of inheritable biological material which is part of a strand of DNA or RNA glycolysis – the biological process of breaking down sugars with the formation of ATP

gradualism – the Darwinian way that evolution has occurred with the appearance of innumerable intermediate life forms as new species have evolved

inorganic – of mineral origin. Not of living origin

ions – electrically charged atoms or groups of atoms. Their electrical charges control their interaction with other ions

invertebrates – animate organisms which lack bony vertebrae and a notochord

kingdoms of life – the most basic groupings of related species. There are three kingdoms of multi-celled organisms and two of single-celled organisms. The multi-celled kingdoms are the animates, plants and fungi and the single-celled kingdoms are the prokaryotes which do not have nuclei and the eukaryotes which have them

labile – changeable

limb girdle – the complex of bones which links either the paired fore limbs or the paired hind limbs to the trunk of a PTV organism

mandible – the jaw bone

morphology – the form of plants and animals

neoDarwinism – the modern form of Darwin's theory incorporating the discoveries of Mendel and other biologists

notochord – the non-bony spinal cord around which bony vertebrae develope. As they do so they mask the embryonic form of the notochord

nucleus – the biological structure which is present in nucleated cells. Within it is concentrated significant amounts of the cell's DNA

nucleated species – single-celled eukaryote species and multi-celled animate, plant and fungal species

organic – of living origin. Not of mineral origin

organisms – complete single-celled or multi-celled individuals photosynthesis – the process whereby plants and some single-celled organisms use the sun's energy to convert carbon dioxide and water into sugars and oxygen

prediction – a prediction is a forecast of the future and technically it is not possible to make prediction about the past. Predication is the correct word to use in these instances but it is an infrequently-used word which may not imply the high degree of accuracy which is inherent in the word "predict"

prebiotic – prior to life. Used to describe organic molecules which are not yet alive

prokaryotes – organisms which lack nuclei

PTV's – pentadactyl tetrapodal vertebrates – five digited four limbed bony vertebrates. All mammals, birds, reptiles and amphibians are PTV's. Fish are the other great group of bony vertebrates and as they lack paired fore and hind limbs, they are not PTV's

radiographs – also called Xrays or roentgenographs

RNA – ribonucleic acid. A biological acid which is closely related to DNA. In some instances it has gene-like characteristics within cells, but its activities are always subservient to the cell's DNA

ribosomes – molecular conglomerates within cells which manufacture the proteins the cell needs for its living activities. The activities of the ribosomes are controlled by RNA

species – singular or plural collection(s) of very similar organisms which can interbreed and produce offspring which are very similar to their immediate antecendents. The offspring can also interbreed with any suitable contemporary organism of the species

taxonomy – the scientific term for classifying species

TMJ's – temporo-mandibular joints – the joints between the base of the skull and the jaw

triangle of study – the method which can be used to study past life. If a feature has been inherited unchanged over a long period of time, the living descendants of the first organism which had the feature will all have the same feature. This means that the study of one example of the feature will provide the same information as will the study of the same feature in any other organism, living or dead, including the common ancestor. It also permits the prediction to be made that if all the living descendants of a common ancestor have the same feature, all the generations of organisms which have been intermediate between the ancestor and its living descendants, also had the same feature in the same form that it has today

vertebrates – animates which have a notochord and vertebrae